Signals and Communication Technology

For further volumes:
http://www.springer.com/series/4748

Raul L. Katz · Taylor A. Berry

Driving Demand for Broadband Networks and Services

 Springer

Raul L. Katz
Taylor A. Berry
Telecom Advisory Services
New York, NY
USA

ISSN 1860-4862   ISSN 1860-4870   (electronic)
ISBN 978-3-319-07196-1   ISBN 978-3-319-07197-8   (eBook)
DOI 10.1007/978-3-319-07197-8
Springer Cham Heidelberg New York Dordrecht London

Library of Congress Control Number: 2014940932

Printed on acid-free paper

Springer is part of Springer Science+Business Media (www.springer.com)

Preface

An earlier version of this study was developed as part of the *info*Dev/World Bank Broadband Strategies Toolkit,[1] with generous funding from the Korea ICT4D Trust Fund. The authors thank the World Bank for authorizing its publication. We are grateful for the comments received while developing the original material, particularly the feedback from the following World Bank professionals: Tim Kelly, Lead ICT Policy Specialist; Carlo Rossotto, Lead ICT Specialist, Regional Coordinator, ECA, and MENA; and Masatake Yamamichi, consultant—ICT Sector Unit. Additional reviewers of the original material included Wonki Min, Ministry of Information and Communication—Republic of Korea; Toru Nakaya and Atsushi Ozu, Ministry of Internal Affairs and Communications—Government of Japan; Phillippa Biggs, Corporate Strategy Division—International Telecommunication Union; Akshaya Sreenivasan, Pennsylvania State University, John DeRidder, Independent Consultant, Australia; Suraj Ramgolam, National Computer Board—Mauritius; John Roman, Director of Broadband and Regulatory policy—Intel's Global Public Policy Group; and Renee Wittemeyer, Director of Social Impact—Intel.

This version contains numerous updates to the original case studies. In addition, whenever possible, data have been updated to reflect the rapidly changing landscape of broadband technology.

A large portion of the material contained in this study was generated throughout the course of multiple consulting engagements supporting the development and implementation of national broadband plans for Costa Rica (*Estrategia Nacional de Banda Ancha*) and Ecuador (*Plan Nacional de Banda Ancha*) as well as digital agendas for Colombia (*Plan Vive Digital*) and Mexico (*Estrategia Digital Nacional*). The authors are grateful for the fruitful collaboration with numerous individuals that occurred throughout the implementation of those projects. In particular, they thank Hannia Vega (former Vice Minister of Telecommunications of Costa Rica), Ana Valdiviezo (Head of Ecuador National Telecommunications Council), Diego Molano (Minister of ICT of Colombia), and Alejandra Lagunes (Coordinator of Mexican National Digital Strategy).

---

[1] The original source material and additional case studies can be retrieved at http://broadbandtoolkit.org/en/home.

Additionally, portions of the work were developed while completing multiple studies for the International Telecommunications Union. The authors acknowledge their work with Nancy Sundberg, Senior Program Officer—Regulatory and Market Environment Division and Youlia Lozanova, Telecommunication—ICT Regulatory Analyst, both from the Telecommunications Development Bureau.

Finally, this study benefitted from numerous collaborations with a number of academics in the field of telecommunications policy. We thank, in particular, Prof. Hernan Galperin, from the Universidad de San Andres (Argentina); Prof. Alison Gillwald, from the University of Capetown (South Africa); Prof. Judith Mariscal, from the Centro de Investigación y Docencia Económica (México); and Prof. Eli Noam, from Columbia University (United States).

Several colleagues at Telecom Advisory Services worked with us in engagements that served as a basis for parts of the work contained in this study. In particular, we thank Dr. Ernesto Flores-Roux, Dr. Pantelis Koutroumpis, and Fernando Callorda.

Contents

Chapter 1
Introduction

This book focuses on the strategies for stimulating broadband demand. The debate around the digital divide has been, so far, driven mainly by statistics based on the number of households that have a fixed broadband connection and a computer, and individuals that have a wireless broadband device, such as a smartphone or tablet. Along these lines, policy emphasis has been made, to a large degree, to increase the deployment of broadband networks (in other words, the supply side). While the causality between network deployment and broadband penetration certainly exists, it is important to consider that a substantial portion of the digital divide is also explained by the demand gap, the reasons for which will be discussed in-depth throughout this book. While the supply gap measures the portion of the population of a given country that cannot access broadband because of lack of service, the demand gap focuses on the potential users that could buy broadband service (since operators offer it in their territory, either through fixed or wireless networks) but do not (see Fig. 1.1).

According to Fig. 1.1, the supply gap is defined by the number of households where either fixed or mobile broadband is not available (bb), while the demand gap is measured by the non-subscribing households of those where broadband is available (dd). Accordingly, the concept of digital divide represents the sum of both groups (bb + dd). While policy discussion has been intense regarding the need for providing universal coverage (and therefore, eliminating the supply gap), the demand gap has not benefitted from an equal level of attention.

Tackling the demand gap is critical for policy-makers, since even in some mature countries it can reach close to 30 % of served households. The research on the social and economic impact of broadband indicates increasing returns to scale derived from enhanced adoption. In other words, the higher the broadband and ICT adoption, the more important the economic and social benefits are.[1] In that sense,

---

[1] For an assessment of returns to scale in ICT, see Roller and Waverman (2001), Koutroumpis, (2009), Katz and Koutroumpis (2013).

R. L. Katz and T. A. Berry, *Driving Demand for Broadband Networks and Services,*
Signals and Communication Technology, DOI: 10.1007/978-3-319-07197-8_1,
© Springer International Publishing Switzerland 2014

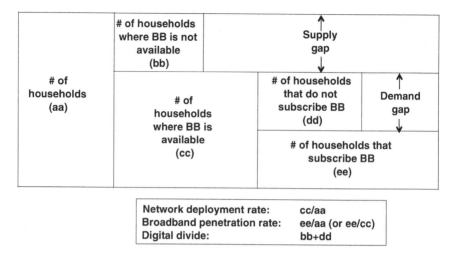

Fig. 1.1 Relationship between supply and demand gap

beyond deploying the necessary infrastructure, stimulating broadband adoption and usage is critical for achieving social development, economic performance, and overall welfare.

The ideal situation for a broadband market is one in which the technology is deployed and service provided by operators on a commercial basis that can be subscribed on a voluntary basis. Because broadband is a basic element of societal infrastructure, it is desirable for operators' business to be sound. Operators will continue to provide broadband service in the long term if they are able to achieve a sustainable profit under sound market mechanisms that can be achieved by government's proper supervision. In order to achieve this win–win relationship between stakeholders of users, operators, and government, an adequate amount of demand has to exist in the market. It is, therefore, necessary for policy-makers to enhance demand with the purpose of maintaining the win–win relationship, without distorting market competition. Also, there may be certain areas in which operators cannot make a profit and do not provide their service. This creates a supply gap, which requires governments to implement policies addressing these market failures.

This book introduces readers to the benefits of higher adoption rates; it examines the progress made so far by countries in the developed and emerging worlds in stimulating broadband demand. It provides an explanation of concepts, such as a supply and demand gap, broadband price elasticity, and demand promotion. In doing so, it also explains differences between fixed and mobile broadband demand gap, introducing the notions of substitution and complementarity between both platforms. Building on these concepts, the study provides a set of recommendations of the best practices, potential strategies, and case studies aimed at promoting broadband demand.

To reiterate, the focus here is not the supply gap, but rather the obstacles for adoption on the demand side. A supply gap is found in places where broadband infrastructure is rudimentary or exhibits limited deployment. This access gap may indeed derive from regulatory policies, but can also come from a lack of interest from investors and telecom operators. Addressing this gap and identifying the common parameters behind it is critical for most national broadband strategies. It is one of the primary purposes of policies aimed at extending universal broadband access and use.

In addition, this book focuses heavily on strategies to affect the behavior of subscribers, whether they are residential users or Small and Medium Enterprises. In this sense, it is less focused on the technological aspects, or even on specific regulatory approaches such as network neutrality. Building on the evidence and the best practices from past programs, the ongoing work in the field of broadband demand stimulation will be described and analyzed to provide a holistic tool and guide regulators and policy-makers. In line with constantly updated information resources, the module has direct links to the sources of information.

References

Katz, R., Koutroumpis, P.: Measuring digitization: a growth and welfare multiplier. Technovation **33**(10), 314–319 (2013)

Koutroumpis, P.: The economic impact of broadband on growth: a simultaneous approach. Telecommun. Policy **33**(9), 471–485 (2009)

Roller, L., Waverman, L.: Telecommunications infrastructure and economic development: A simultaneous approach. Am. Econ. Rev. 909–923 (2001)

Chapter 2
Assessing the Broadband Demand Gap

Before implementing specific demand stimulation strategies, policy-makers must conduct a diagnostic that allows them to determine the size and sources of the demand gap. This process begins by estimating the percentage of the population that can purchase broadband, yet still do not. Once this gap is quantified, it is necessary to understand the drivers of this so-called "failure." Is it because a portion of the population cannot afford to purchase a subscription at current prices? Or is it because they lack the necessary digital literacy that allows them to access the Internet? It could also be the case that while potential users have a computer (or comparable device), they cannot find any online content, applications, or services that would motivate them to purchase broadband service.[1]

This chapter explains the different concepts and provides examples of methodologies for measuring the demand gap and constructing a diagnostic of structural factors affecting adoption. It would set the stage to explain a variety of approaches and policy solutions to meet the adoption targets.

2.1 Measuring the Broadband Demand Gap

Measuring the demand gap is the first step in the development of a diagnostic that will lead to the formulation of demand stimulation policies. Given the interrelationship of fixed and mobile broadband leading to complementarity and/or substitution scenarios, this exercise is not trivial. Moreover, measuring demand gaps in the aggregate for a whole country is not necessarily a suitable approach for the development of targeted policies. Therefore, any attempt at measuring the broadband demand gap has to be conducted at a disaggregated level (county, department). This section first addresses how to measure the demand gap in fixed broadband, then moves to the mobile broadband gap, and finally discusses the interrelationship between both domains.

---

[1] While one of these three reasons could be found to be the dominant one in certain population groups, the likely scenario is one of high correlation between the three variables.

R. L. Katz and T. A. Berry, *Driving Demand for Broadband Networks and Services,*
Signals and Communication Technology, DOI: 10.1007/978-3-319-07197-8_2,
© Springer International Publishing Switzerland 2014

2.1.1 The Fixed Broadband Demand Gap

Demand gap is defined as the difference between either households or individuals that could gain access to broadband but do not acquire the service. This is not a statistic that is typically being tracked by either regulators or made public by operators. In recent years, however, policy makers, driven by the need to develop national broadband strategies and plans, have in some instances been able to estimate this metric.

While most countries have fairly accurate estimates of broadband subscribers, they lack a solid grasp of network coverage, defined as the proportion of the population of a given country that is "served" by broadband technology. This metric (and the supporting coverage maps) should be calculated both for fixed and mobile broadband.

In the case of fixed broadband, coverage needs to be estimated in terms of the number of households that are served by broadband providers (i.e., where residents have the option to purchase service from telecommunications carriers, cable TV operators, or fixed wireless providers such as WiMax). Even this number can be sometimes difficult to estimate. For example, the development of the United States' National Broadband Plan introduced the notion of the "underserved" household. "Underserved" means that the resident can get access to broadband, but at a download speed below the target stipulated by the broadband plan (in this case, 2 Mbps). Therefore, a first level assessment should consider three categories of fixed broadband coverage: "served," "underserved" (download speeds lower than the target), and "unserved" (no service at all). The problem with the "underserved" category is that in emerging countries, a large portion of households can gain access to service download speeds much lower than those stipulated in a broadband plan (for example, 256 kbps). Given the hurdle to improve the level of service to the "underserved" population, the general consensus is that for the time being, at least in emerging countries; attempting to reach mass-deployment levels of broadband, this category should not be considered as part of the estimation of the broadband demand gap.

Another difficulty in assessing fixed broadband coverage resides in the interpretation of operator-provided information. The introduction of certain modifications to existing telecommunications and cable TV networks enables broadband deployment in its most basic mode. In the case of telecommunications copper networks, xDSL service requires the installation of equipment at the central office, while in the case of cable TV, cable modem service requires the upgrading of its networks to bi-directional 750 MHz capacity. The implication of this situation is that a residence could have either wireline telephony or cable TV coverage, but the infrastructure is not upgraded to the point where it may have the capability of handling a subscriber's request for service. The question of interpretation, then, is whether that residence should be considered "served" or "unserved". It is generally accepted that, in the case of emerging countries, if a telecommunications

Table 2.1 Developed countries: fixed broadband demand gap (2011)

Country	Households covered (%)	Households connected (%)	Demand gap (%)
Australia	89	69	20
Denmark	96	76	20
France	100	77	23
Germany	98	58	40
Israel	100	83	17
Italy	95	55	40
Korea, Rep.	100	93	7
Spain	93	61	32
Sweden	100	89	11
United Kingdom	100	68	32
United States	96	64	32

Sources Katz and Galperin (2012) based on ITU data

fixed network or a cable TV system serves the customer, he/she should be included in the "served population" category.

With these two caveats in mind, the demand gap can be calculated by using a standard coverage metric estimation. For example, Table 2.1 presents data on the fixed broadband demand gap for select developed countries.

As shown in Table 2.1, the broadband demand gap is not only an emerging market phenomenon. In certain developed countries (such as Germany, Spain, Italy, the United Kingdom, and the United States), an important portion of households lack broadband connectivity for reasons other than service availability.

In the United States already in 2009,[2] for example, 96 % of households were served by cable modem technology, while 82 % could acquire broadband service from the telecommunications operator. However, as indicated in the statistics of Table 2.1 in 2011, only 64 % of households purchased service. Therefore, 32 % of households could have access to broadband services, but choose not to acquire a subscription. As expected, the demand gap in this country varies by state as shown in Fig. 2.1.

As Fig. 2.1 indicates, the broadband demand gap is larger in less-developed states. In Mississippi, for example, it is 60 %, while the supply gap (non-served households) is 9 %. In a more economically-developed state such as Georgia, where service penetration is higher, the supply gap is 8 %, while the demand gap is 34 %.

In a European country such as Germany, according to the National Broadband Strategy published in February of 2009, 98 % of all households (39,700,00) could already access broadband service. Of these, 37,600,000 could be served by xDSL, 22,000,000 were served by cable TV (and therefore could buy broadband via cable

---

[2] Rather than providing the latest statistics, the purpose of the following examples is to demonstrate how to calculate the demand gap and provide a comparison among countries based on orders of magnitude.

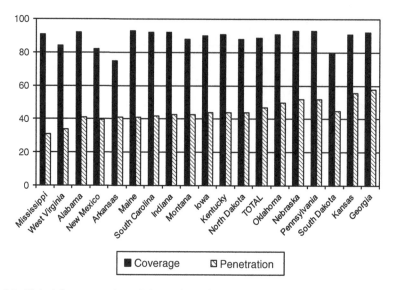

Fig. 2.1 United States: supply and demand gap for states with lowest broadband penetration (percent of households) (2009). *Source* FCC (Table 14 of HSPD1201); US census Bureau

modem), and 730,000 could access broadband via fixed wireless or satellite. However, despite the near-complete coverage, only 58 % of households purchased broadband, signifying a demand gap of 40 %.

The fixed broadband demand gap is, as expected, a more serious problem in emerging countries. Table 2.2 presents statistics for coverage and demand for Latin America countries.

Latin America displays an average demand gap of 43 %, which means that less than half of covered households are purchasing broadband subscriptions. As Table 2.2 indicates, more extensive coverage results in a higher demand gap. These metrics, typical of emerging countries, indicate that supply in Latin America does not appear to be the dominant hurdle to increasing broadband penetration. They demonstrate, rather, the criticality of demand stimulation strategies targeting either the affordability or the awareness structural factors.

2.1.2 The Mobile Broadband Demand Gap

Measuring the demand gap in mobile broadband presents methodological problems as in the case of fixed broadband networks. First, it is generally agreed that policy makers should consider at least 3G networks to be the technology benchmark when measuring mobile broadband coverage. While 3.5G, HSPA, or LTE networks are the obvious platforms to provide a relatively smooth Internet access experience, in

Table 2.2 Latin American countries: fixed broadband demand gap (4Q2012)

Country	Coverage (%)	Household penetration (%)	Demand gap (%)
Argentina	96	40	56
Bolivia	40	4	36
Brazil	94	34	60
Chile	78	50	28
Colombia	81	32	49
Costa Rica	95	36	59
Ecuador	80	26	54
Mexico	62	53	9
Peru	59	20	39
Average	76	33	43

Sources for coverage Katz and Galperin (2012); penetration based on ITU

emerging country contexts, it is advisable to measure coverage once again at a slightly lower speed, such as the one comprised by the whole WCDMA family.[3]

Secondly, the concept equivalent to the "underserved" category in fixed broadband exists in the case of mobile broadband as well: in this case, it is labeled the "gray" zones. These represent the areas covered by wireless networks affected by either capacity or signal propagation limitations. Again, while some national broadband plans have been very emphatic about measuring these zones (see Germany's National Broadband Strategy), in the case of emerging countries, it might be convenient to set this measurement aside for the next few years.

Thirdly, mobile broadband adoption needs to consider the device utilized to access the Internet. The first category of devices includes, quite naturally, all modems that can be relied upon to access the Internet from a PC, a laptop, or a netbook. These devices include dongles, USB modems, and air cards. The integrated devices such as tablets, and smartphones that provide adequate screen formats and interface to surf the web, respond to emails, and access common web platforms such as Google, YouTube, or Facebook comprise the second category. This category would exclude feature phones, which, by virtue of their small screen formats and keyboards, have limited broadband access ability. For example, the pioneering work of Horrigan (2012) on the value of mobile broadband to close the digital divide in the state of Illinois focuses only on smartphone adoption.

In light of these issues, how should mobile broadband coverage and adopters be measured? Beyond shipment statistics and installed base for selected operators, the number of subscribers that own an Internet suitable device connected to a 3G or higher performance network is not readily accessible. On the other hand, the

---

[3] For example, when Japan implemented its (1) New IT Reform Strategy which was set in 2006 by IT Strategic Headquarters (headed by the Prime Minister) and (2) the Digital Divide Elimination Strategy which was set in 2008 by the Ministry of Internal Affairs—both of which aimed to eliminate all broadband zero areas by the end of FY2010 (March 2011), 3.5G was considered the minimal broadband service.

Table 2.3 Developed countries: mobile broadband demand gap (2011)

Country	Population covered (%)	Population connected (%)	Demand gap (%)
Australia	97	89.10	7.9
Denmark	97	57.51	39.49
France	98.20	32.86	65.34
Germany	86	34.76	51.24
Israel	99	54.36	44.64
Italy	91.86	48.19	43.70
Korea, Rep.	99	97.13	1.87
Spain	90.60	36.68	53.92
Sweden	99	85.10	13.90
United Kingdom	95	42.56	52.44
United States	98.50	71.91	26.59

Source Katz and Galperin (2012)

number of 3G and 4G subscribers is easier to access. Therefore, it would be advisable to gather those statistics to measure the mobile broadband demand gap. Mobile broadband coverage should be measured in terms of 3G coverage, a metric provided by either the ITU or commercially available databases such as GSMA Intelligence. However, the estimates provided by these sites are only presented at the national level, preventing a detailed regional analysis.

Table 2.3 presents statistics on mobile broadband demand gap for selected developed countries.

As the Table 2.3 indicates, with a few exceptions (Australia, Republic of Korea, United States), the mobile broadband demand gap of countries studied is higher than the fixed broadband demand gap. These numbers should be interpreted with the caveat that the latter measures the household gap while the former measures population.

In the case of emerging countries such as those of the Latin America region, the mobile broadband demand gap is even higher (see Table 2.4).

As Table 2.4 indicates, the average mobile broadband demand gap in Latin America is 57 %, which means that 57 % of the Latin American population could purchase a mobile broadband connection but do not. This difference requires an analysis of the obstacles faced by users to acquire broadband service. An understanding of such factors will allow policy makers to deploy the relevant initiatives to tackle these obstacles. This is addressed in Chap. 3, 4 and 5 below.

2.1.3 Demand Gap and the Interrelationship Between Fixed and Mobile Broadband

Until now, we have treated the demand gap within fixed and mobile broadband as two independent phenomena. This treatment is somewhat artificial since both technologies are offered within adopters' same universe. Naturally, each platform

Table 2.4 Latin American countries: mobile broadband demand gap (2012)

Country	Population covered (%)	Population connected (%)	Demand gap (%)
Argentina	92	21.87	70.13
Bolivia	29	6.92	22.08
Brazil	84	32.83	51.17
Chile	82	27.04	54.96
Colombia	96	8.69	87.31
Costa Rica	93	36.22	56.78
Ecuador	86	21.92	64.08
Mexico	77	20.63	56.37
Peru	63	11.70	51.30
Average	78	20.86	57.14

Fuentes For coverage, Katz and Galperin (2012); penetration based on GSMA Intelligence

meets specific requirements. Mobile broadband adds the mobility premium to the Internet access experience. At the same time, at least for now, due to given technology and shared resource limitations, mobile broadband networks are not the most suitable platform to fulfill certain applications, like downloading movies or playing massive parallel games. This factor notwithstanding, it is generally assumed that, given their ease of deployment, mobile broadband networks are very appropriate to fulfill coverage requirements in emerging countries. If that were to be the case, how should policy makers think about the interplay between both platforms?

At the initial steps of diffusion processes, mobile broadband technology represents a complementary technology to fixed broadband. The early adopter of mobile broadband is, most likely, already a subscriber of fixed broadband. In this situation, mobile broadband complements fixed broadband by providing the added value of mobility. An example of this situation is that of Mexico (see Fig. 2.2).

As the Mexican example indicates, mobile broadband subscribers through the end of 2010 were likely already fixed broadband customers belonging to high socio-demographic segments for which mobile broadband represented an added value proposition to meet their Internet connectivity needs.

However, the complementarity consumption pattern is not the only trend. In many cases, especially in emerging countries, mobile broadband represents a substitute to fixed broadband. This occurs under three possible situations: (1) when the fixed broadband service is not being offered in the area where the customer resides, (2) when the quality of fixed service is at a disadvantage with respect to the mobile offering (for example, in terms of speed), or (3) when the user decides to consolidate services to reduce expenditures and acquires the mobile service that provides both connectivity and mobility. It is important to mention that the last situation can occur in the context where the applications and services to be accessed are interchangeable between the two platforms. While this is possible, as it was mentioned earlier, there are services that are better suited to the fixed technology and cannot be fully accessed by mobile broadband. Additionally, a

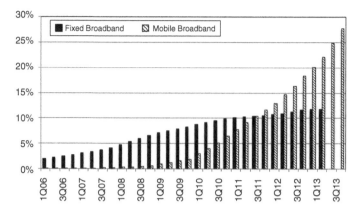

Fig. 2.2 Mexico: penetration of fixed versus mobile broadband (2006–2013). *Source* Katz (2012)

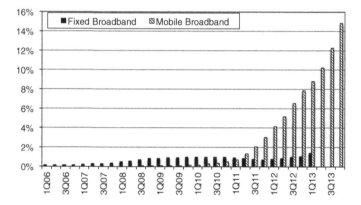

Fig. 2.3 Bolivia: penetration of fixed versus mobile broadband (2006–2013). *Source* Katz (2012)

limiting factor of mobile broadband's substitution power is the prevalent pricing plans that institute caps on the amount of data subscribers can download on a monthly basis. However, as will be shown in Chap. 4, capped mobile broadband offerings are highly suited to deliver broadband service to population at the bottom of the socio-demographic pyramid.

The case of Bolivia is a good example of substitution of fixed broadband by mobile broadband service (see Fig. 2.3).

Bolivia is a country that arrived fairly late to the Internet revolution. At the outset of mobile broadband in the country, the adoption of the fixed platform had not reached 1 %, a common scenario in many emerging countries. Not surprisingly, the high price of the offering contributed to this limited adoption. In the second quarter of 2010, the least expensive plan for a 2.5 Mbps download speed cost the equivalent of US$ 325/month. That year, a wireless service providing broadband connectivity to a PC via a USB modem was priced at the equivalent of US$ 16.38. The cross-elasticity of both offerings resulted in a decline in the number of fixed broadband lines and a dramatic increase in mobile broadband accesses, a classic case of substitution. The Bolivian example could be quite applicable to the experience of other emerging countries.

In considering the substitution scenario to understand the broadband demand gap, it is important to include in the broadband subscribers numbers those users who purchase only mobile broadband service and add them to the customers who have acquired fixed broadband. Therefore, the broadband demand gap that considers mobile broadband subscriptions has to be quantified according to the following formula:

Broadband Demand Gap = Broadband Coverage (C) – Broadband Subscriptions (S)

C = Population covered by fixed and mobile broadband + Population covered only by fixed broadband + Population covered only by mobile broadband

S = Subscribers of fixed and mobile broadband (complementarity) + Subscribers of only fixed broadband + Subscribers of only mobile broadband (substitution)

According to this formula, the precise estimation of the broadband demand gap requires a solid understanding of parameters such as the degree of complementarity of fixed and mobile platforms, to avoid double counting of mobile and fixed subscribers. Unfortunately, this type of statistic does not exist for the time being in most countries, which obliges the policy maker to continue relying on demand gap assessment by technology. Some countries are currently conducting field research aimed at evaluating the degree to which both technologies complement or substitute each other. As was discussed earlier, the answer will depend on factors such as geography and socio-demographic segment. The first results of such a research are being generated primarily in developed countries. Horrigan (2012) published the results of survey research conducted in the state of Illinois testing the hypothesis of mobile broadband contributing to closing the digital divide. The results indicated the existence of only a small portion of households accessing the Internet through mobile platforms: of 86 % of at-home broadband subscribers, 53 % had fixed broadband and smartphones, 26 % had only fixed broadband, and only 7 % had only smartphones. The interesting finding is that the 7 % of subscribers that accessed the Internet only through smartphones tended to have a low income (below $20,000), lower education than the general population and lived in rural areas. While the activities these users conduct on the Internet are less data intensive than those that have both fixed and mobile broadband, one could argue that the latter contributes to some extent to close the digital divide.

A similar conclusion was formulated by Bohlin et al. (2011), who estimated in their study of Sweden broadband that the probability of using mobile broadband exclusively tends to increase if the respondent is aged 35 years or less, and has low income.

The primary caveat is that both studies were conducted in an environment of widely diffused fixed broadband. In emerging country settings, with less fixed broadband coverage, the substitution power of mobile broadband is much larger.

2.2 Understanding the Residential Broadband Demand Gap

Once the broadband demand gap is quantified, policy makers need to understand the factors driving that gap. In this section, the obstacles and drivers of the residential demand gap will be first reviewed. At its conclusion, the same approach will be followed for the enterprise side, focusing primarily on small and medium enterprises.

To reiterate, the primary topic in this section is the adoption gap. An adoption gap is found in places where broadband infrastructure is in place but often underutilized. This is a typical low demand case that reflects a low desirability for the services offered or a relatively high cost of ownership. The broadband demand gap can be the result of multiple factors. In fact, the obstacles could be different by region of the country, and by socio-demographic group.

Research on the variables affecting broadband diffusion is quite extensive. For example, Hauge and Prieger (2010) point out that income, educational level of the head of household, and household age composition are the main predictors of broadband adoption. Other studies mention variables that are more specific to countries or regions. Navarro and Sanchez (2011) indicate that, *ceteris paribus*, gender is a strong predictor in Latin America, where females are 6 % less likely to adopt broadband. In the United States, several studies underscore the importance of factors such as ethnicity and mastery of the English language (Ono and Zavodny 2007; NTIA 2011). Other factors such as the location of potential subscribers (rural versus urban), the presence of school children in the household (Horrigan 2014), and the penetration of broadband in the location where the potential adopter resides are also important factors driving Internet adoption (see Chaudhuri et al. 2005; Vicente and Lopez 2006; Grazzi and Vergara 2011).

At the highest level of analysis, the residential broadband demand gap is the result of three obstacles:

- Limited affordability: certain portions of the population either cannot acquire a device or purchase the subscription needed to access the Internet
- Limited awareness of the potential of the service or lack of digital literacy
- Lack of relevance or interest: the value proposition of applications, services, and content does not fulfill a need of the adopting population

Each of these three obstacles are driven by one or a combination of four structural variables:

- Income levels: the socio-demographic group, measured by income, does not only influence the affordability barrier, but is also correlated with limited awareness and lack of relevance
- Education levels: the education attained by the potential user influences the degree of digital literacy and is related to interest in accessing the Internet
- Age: similarly, the age variable is inversely related to digital literacy and content relevance
- Ethnicity: as a result of linguistic and/or cultural structural factors, ethnic group belonging can impact the level of interest in accessing the Internet

These relationships have been depicted in Fig. 2.4.

The research literature has also studied the role played by other structural variables, such as gender. For example, a gender gap was detected in some Latin American countries (see Universidad Alberto Hurtado 2009 for Chile; INEI 2012 for Peru; and Rectoría de Telecomunicaciones 2011 for Costa Rica). Our research in Ecuador (Katz and Callorda 2013) found that in townships where broadband was installed, the average income of women in households with no computer or Internet usage was lower than that of men two years later. However, if the household had a computer or the family accessed the Internet prior to broadband deployment, the effect on average income was similar for men and women. This would indicate that a gender gap does exist at lower socio-demographic levels. In support of this finding, research by Hilbert (2011) concluded that the gender gap disappears when control variables such as income and education are included in the analysis. Beyond the importance of socio-demographic level as the primary explanatory variable, the maturity of the technology also helps erasing gender differences. For example, research conducted in Asia by the Korea Network Information Center (KANIC) showed that the gender composition of Internet users has shifted toward equality from 33 % female in 1999 to 45 % female in 2002. These findings should be taken into consideration when examining the viability of gender-based policy initiatives for having a positive contribution to stimulating adoption (these will be discussed in detail in Sect. 3.12).[4]

The following section will explain each of the three obstacles—affordability, awareness, and relevance—and link them back to the structural variables. In each section, studies and data regarding the obstacles and driving variables in developed and emerging countries are presented.

---

[4] While the gender issue is negligible as one of the structural variables within this analysis, the gender divide still needs to be addressed. Please see "Digital literacy for Women" section (currently, Page 65 of the World development Report).

Fig. 2.4 Broadband
adoption structural factors

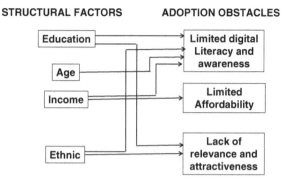

2.2.1 The Affordability Barrier

The economic barrier remains a key factor in limiting broadband adoption. However, it would seem that in developed countries with higher household incomes, the economic barrier takes second seat to either low digital literacy or cultural inadequacy.

In the United States, broadband non-adopters amount to 22 % of households (34.6 million). Within this group, the head of household is older than 65 years old (30 %), has not completed high school studies, belongs to a household with less than US$ 30,000 income (41 %), and has a limited understanding of the English language (Pew Center, 2012). In a study by the National Telecommunications and Information Agency (NTIA 2011) researching the reasons of non-adoption, 24 % of respondents mentioned affordability of either devices or service. If the household already had a PC, 37 % of respondents mentioned broadband affordability as a barrier. Conversely, the households that did not have either a PC or broadband amounted to 21 %. Consequently, in the United States, if a household already has a computer, broadband affordability becomes an important adoption barrier. The Pew Research Center, in a research similar to that of the NTIA, found that 19 % cite cost which was made up of 13 % saying they do not have a computer and 6 % saying it is too expensive.

In a statistic close to that of the United States, a 2009 survey among Australia's non-adopters showed cost to be the most common deterrent cited amongst this group: 26 % of respondents reported that it was too expensive (Australian Government Information Management Office 2009).

In Spain, a study (ONTSI 2011) indicated that of the 7 million non-adopting households, 42 % indicated affordability as a primary barrier to adoption. The distribution of this population is, not surprisingly, tilted towards households of the lower socio-demographic segments. For example, 52 % of households with monthly income of less than 1,100 Euros mention affordability as a primary barrier; that percentage decreases to 16 % among households with incomes higher than 2,700 Euros.

In the United Kingdom, of the 24 % of non-adopting households, only 16 % mentioned the affordability barrier, while 66 % mentioned lack of perceived content relevance. In a similar study conducted in 2010, the percentage citing cost as an obstacle to adoption was 23 %. This statistic suggests that as prices decrease, digital literacy and relevance structural factors (to be discussed in the next section) have more of an impact.

Moving now to emerging countries, the importance of affordability increases among the reasons mentioned for non-adoption. For example, in a household ICT survey conducted in Brazil, 48 % of non-adopters of broadband mentioned cost as being the dominant reason (CGI 2012). In research conducted as part of the National Household Survey, 60 % of Mexican households with no computer and no broadband connection mentioned cost of access as the dominant reason (INEGI 2011). In Chile, the number was slightly less: 37 % (Subtel 2009), while in Costa Rica, it was 60 % (Rectoría de Telecomunicaciones 2011). The percentage mentioning affordability as the dominant reason in Chile was quite close to households surveyed in Colombia: 39.9 % (MITIC 2011). In Puerto Rico, 16 % of non-adopters mentioned price as being too high (Puerto Rico Broadband Task Force 2012).

Research conducted in Sub-Saharan Africa indicates that while availability does play a role in this difference in uptake, the cost of broadband services in the region is prohibitively high. In 2006, the average price for basic broadband in Sub-Saharan Africa was US$ 366 per month, when in India the average rate ran from $6 to $44 per month (Williams 2010).[5] In much of Europe, these rates ranged from $12 to $40 per month.

Not surprisingly, a 2011 study demonstrated that many of the countries with low broadband adoption rates were the same countries with the highest annual costs for broadband per gross national income (Point Topic, 2011).[6] This list of countries included Kenya, where the average annual cost for broadband amounted to 79.25 % of gross national income. In comparison, the cost of broadband access in Switzerland—which ranks amongst the top 10 countries in the world in terms of broadband penetration—amounted to a mere 0.07 % of gross national income.

A compilation of all the statistics reviewed earlier indicates that affordability remains a preeminent variable in explaining the non-adoption of broadband, particularly in emerging countries (see Table 2.5).

As the data in Table 2.5 suggests, the lower the level of disposable income, the higher the importance of the affordability barrier becomes. This conclusion is supported as well by the penetration of broadband by decile of the socio-demographic pyramid.

---

[5] https://openknowledge.worldbank.org/bitstream/handle/10986/2422/536430PUB0Broa101 Official0Use0Only1.pdf?sequence=1

[6] http://broadband.about.com/od/barrierstoadoption/a/Affordability-As-A-Barrier-To-Broad band-Adoption.htm

Table 2.5 Percentage of households mentioning affordability as reason for not purchasing broadband

Country	Percentage (%)	Source
United States	24	NTIA (2011)
United States	36	Horrigan (2013)
United Kingdom	16	OFCOM (2012)
Spain	42	ONTSI (2011)
Australia	26	AGIMO (2009)
Chile	37	Subtel (2009)
Brazil	48	CGI (2012)
Colombia	40	MITIC (2011)
Costa Rica	60	Rectoría de Telecomunicaciones (2011)
Mexico	60	INEGI (2011)
Puerto Rico	16	PRBT (2012)

The statistics of Table 2.6 confirm the disparity regarding broadband adoption between the top three deciles and the three bottom deciles of the socio-demographic pyramid. While the difference between the bottom and top of the socio-demographic pyramid is smaller in emerging countries, that is due to the overall lower penetration of broadband in these economies (see Table 2.7).

While still present in developed nations, it is reasonable to conclude that the affordability barrier to broadband adoption at the bottom of the pyramid is a phenomenon more prevalent in emerging countries (see Fig. 2.5).

As observed in Fig. 2.5, an exponential relationship exists between income per capita purchasing power parity in US dollars (horizontal access) and broadband penetration at the bottom of the pyramid (vertical access). When per capita income of a country surpasses US$ 20,000, fixed broadband adoption at the base of the pyramid exceeds 20 %. This situation confirms that higher income, to a large extent, largely solves the affordability problem.

This conclusion for emerging economies is further supported in studies conducted at the country level. For example, in research conducted by the Costa Rican government, in households with a monthly income higher than 750,000 Colones (local currency), computer and fixed broadband adoption exceeds 80 %; in households with incomes lower than 750,000 Colones, the adoption of both technologies decreases to 60 % and under (see Fig. 2.6).

In another study conducted in Chile, broadband adoption in the highest quintile reached 73 % in 2009, while within the lowest quintile, adoption is 10 % (see Fig. 2.7).

Similar results were obtained in a research conducted in Colombia, where broadband penetration in top social strata was 83 %, while adoption in the bottom was only 2 % (see Fig. 2.8).

Table 2.6 Broadband penetration at bottom 3 deciles versus top 3 deciles of socio-demographic pyramid (2012)

Countries	Broadband penetration (bottom 3 deciles)	Broadband penetration (top 3 deciles)	Countries	Broadband penetration (bottom 3 deciles)	Broadband penetration (top 3 deciles)
Algeria	0.37	12.90	France	33.60	80.83
Argentina	4.43	41.73	Georgia	0.67	7.43
Australia	50.23	91.67	Germany	39.23	91.83
Austria	37.27	83.40	Greece	15.20	65.90
Azerbaijan	0.17	3.63	Guatemala	0.70	5.47
Bahrain	16.67	79.10	Hong Kong, China	46.93	97.87
Belarus	3.50	45.67	Hungary	24.70	75.27
Belgium	39.77	92.57	India	0.33	12.67
Bolivia	1.13	21.83	Indonesia	0.07	4.60
Bosnia and Herz.	5.47	18.83	Iran	1.80	17.47
Brazil	4.63	50.07	Ireland	29.93	76.97
Bulgaria	7.70	47.77	Israel	29.73	72.67
Cameroon	0.13	2.33	Italy	22.13	72.57
Canada	47.03	91.03	Japan	57.10	65.93
Chile	12.73	41.27	Jordan	4.27	17.53
China	7.87	40.20	Kazakhstan	0.13	13.53
Colombia	4.37	26.47	Kenya	0.20	4.70
Costa Rica	5.83	39.17	Korea, Rep.	92.03	100.00
Croatia	14.97	76.17	Kuwait	11.87	60.00
Czech Republic	17.80	84.23	Latvia	27.33	73.43
Denmark	65.47	89.83	Lithuania	28.87	70.83
Dominican Rep.	1.37	11.13	Macedonia	4.57	35.13
Ecuador	0.67	21.37	Malaysia	9.77	45.43
Egypt, Arab Rep.	0.70	11.77	Mexico	4.33	46.10
Estonia	30.63	88.20	Montenegro	7.00	40.87
Finland	51.80	98.13	Morocco	1.33	15.43
Netherlands	71.60	86.17	Norway	74.40	89.47
Nigeria	0.10	12.23	Peru	0.63	25.07
Poland	25.97	83.20	Switzerland	53.37	94.03
Qatar	25.80	100.00	Tunisia	0.47	8.37
Romania	6.27	38.97	Turkey	13.57	54.27
Russia	10.10	46.37	Turkmenistan	0.00	0.63
Saudi Arabia	18.43	63.83	Ukraine	1.40	21.57
Serbia	8.17	48.47	United Arab Emirat.	29.03	100.00
Singapore	60.57	96.80	United Kingdom	41.10	97.10
Slovak Republic	17.97	73.93	United States	49.57	83.17

(continued)

Table 2.6 (continued)

Countries	Broadband penetration (bottom 3 deciles)	Broadband penetration (top 3 deciles)	Countries	Broadband penetration (bottom 3 deciles)	Broadband penetration (top 3 deciles)
Slovenia	31.53	85.93	Uruguay	4.10	38.60
South Africa	0.57	11.97	Uzbekistan	0.00	4.33
Spain	30.30	80.60	Venezuela	5.47	45.37
Sweden	62.20	95.77	Vietnam	0.30	20.50

Source Euromonitor (2012)

Table 2.7 Weighted average penetration in the bottom 3 deciles versus top 3 deciles of socio-demographic pyramid (2012)

	Broadband penetration (bottom 3 deciles) (%)	Broadband penetration (top 3 deciles) (%)	Difference (in percentage points)
Developed countries	38.83	84.60	45.77
Emerging countries	4.80	27.81	23.01
Mean	19.39	52.15	32.76

Source Euromonitor (2011); calculated by the authors

Fig. 2.5 Relationship between level of economic development and broadband adoption at the bottom of the pyramid (2011). *Source* Katz and Callorda (2013) based on Euromonitor and IMF

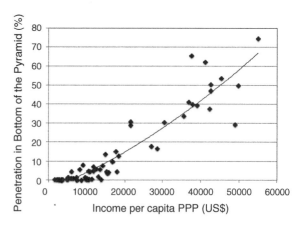

Since the penetration in the bottom of the socio-demographic pyramid is highly correlated with the average household income and the GDP per capita of a given country, it is expected that nations that undergo rapid economic growth or implement poverty-reduction programs would witness a reduction in the broadband affordability gap. Brazil is a clear example of this effect (see Fig. 2.9).

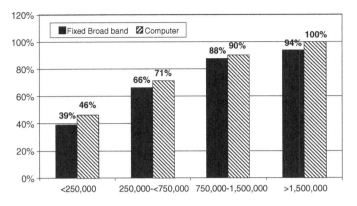

Fig. 2.6 Costa Rica: household computer and fixed broadband access by income (2010). *Source* Costa Rica. Rectoria de Telecomunicaciones

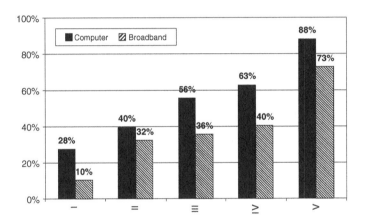

Fig. 2.7 Chile: household computer and broadband access by socio-demographic quintile (2009). *Source* Universidad Alberto Hurtado

Fig. 2.8 Colombia: household computer and broadband access by socio-demographic strata (2010). *Source* SUI; National Department of Statistics

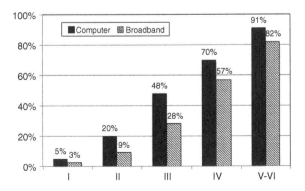

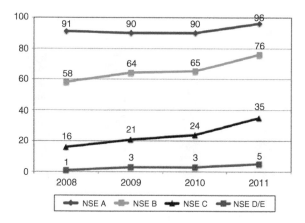

Fig. 2.9 Brazil: household adoption of broadband internet (by socio-economic segment) (2008–2011) *Source* CGI Household Survey. *Note* The classification of socioeconomic level in Brazil is based on the level of instruction and ownership of goods per the Criteria of Economic Classification Brazil (CCEB), which is a part of the Associação brasileira de empresas de pesquisa". It is based on household presence of color TV (0–4 points), radio (0–4 points), bathroom (0–7 points), automobile (0–9 points), maid service (0–4 points), washing machine (0–2 points), DVD (0–2 points), refrigerator (0–4 points), and freezer (0–2 points). On the other hand, the head of the household's level of instruction is also considered depending on whether he or she has a complete higher education (8 points), incomplete higher education (4 points), complete secondary education (2 points), complete primary education (1 point), or if they are illiterate (0 points). In the event that the total number of points is greater than or equal to 35, the household falls into category NSE A; if it totals between 23 and 34 points, category NSE B; between 14 and 22 points, NSE C; between 8 and 13 points, NSE D; and if the points total between 0 and 7 points, they fall into category NSE E

As demonstrated in Fig. 2.9, the increase of fixed broadband adoption in Brazil between 2008 and 2011 is essentially a phenomenon of the upper and middle classes. The figure shows that the penetration jump in the B and C segments could be linked primarily to the income redistribution policies put in place by the administrations of presidents Lula and Rousseff. Segments D/E have not increased their adoption of fixed broadband over time in any significant way, because, the social policies put in place cannot break through the broadband affordability barrier at the bottom of the socio-demographic pyramid.

However, if the economy does not grow or no poverty reduction programs have been actively implemented, the affordability broadband gap tends to increase. For example, in Mexico the increase in broadband penetration has been significant within the higher deciles (VIII to X) between 2008 and 2010 while rates stagnated amongst the lower tiers, therefore accentuating the socio-demographic digital divide (see Fig. 2.10).

The evidence collected both at the aggregate and country level confirms the importance of the affordability variable in explaining a substantial portion of the broadband demand gap. In this context, two policy levers are particularly relevant to affect this dimension of the digital divide. At a macro-economic level, all

Fig. 2.10 Mexico: household internet adoption (by income decile) (2008–2010). *Source* INEGI (2011)

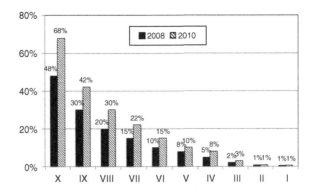

Table 2.8 Latin America: wireless penetration at the bottom of the pyramid (three lowest deciles)

Country	2007	2008	2009	2010	2011	2012
Argentina	43.27	48.30	54.10	58.97	62.90	65.93
Bolivia	21.10	27.50	38.23	49.37	59.97	69.37
Brazil	47.37	57.43	62.50	66.33	68.90	70.80
Chile	49.80	57.53	69.97	78.07	82.07	85.47
Colombia	51.50	64.20	68.13	73.47	76.63	78.87
Costa Rica	35.13	44.17	45.80	51.00	55.27	58.20
Ecuador	35.87	42.27	49.37	54.27	59.40	63.80
Guatemala	35.97	37.17	38.47	38.97	39.80	40.53
México	39.13	46.70	56.50	60.97	64.80	68.03
Peru	13.97	17.77	22.20	26.90	31.90	36.77
R. Dominicana	34.10	40.90	46.67	50.43	53.83	56.97
Uruguay	45.17	55.13	63.90	68.90	73.07	77.80
Venezuela	14.80	15.20	15.57	15.90	16.20	16.57
Total	42.92	48.47	53.34	53.35	57.29	60.70

Source Euromonitor (2012); compiled by Katz and Callorda (2013)

programs aimed at reducing poverty levels (such as the ones implemented in many countries of Latin America, such as Argentina, Brazil, and Venezuela) will, without doubt, have an impact in reducing the demand gap. At the ICT sector level, policies aimed at reducing the pricing of services will necessarily stimulate adoption. The types of programs that could be put in place and their expected impact will be reviewed in Chap. 4.

It should be mentioned, however, that remedies for this market failure—the affordability barrier—should be sought out not only in the area of state intervention in the fixed broadband space, but also through alternative technologies like mobile broadband. In this sense, the potential of mobile broadband to tackle the affordability barrier at the bottom of the pyramid merely replicates the experience of wireless in addressing the universalization challenge of the voice telephony. Table 2.8 illustrates how mobile telephony has been gradually penetrating the bottom of the socio-demographic pyramid in the Latin American region.

As seen in Table 2.8, the combination of pre-paid offers and policies of "calling party pays" has enabled voice telephony penetration rates to reach, on average, 60.70 % at the bottom of the socio-demographic pyramid in 2012, up from 42.92 % in 2007. In this sense, the question is whether or not the mobile platform can replicate the universalization success achieved in voice telephony to broadband, but how it could be done. We will come back to this in Chap. 4.

2.2.2 Limited Digital Literacy

Beyond the affordability barrier, lack of digital literacy can explain a portion of the broadband demand gap. Digital literacy is the ability to navigate, evaluate, and create information effectively and critically using a range of digital technologies. Digital literacy encompasses all devices, such as computer hardware, software, the Internet, and cell phones. Research around digital literacy is concerned not just with being literate at using a computer, but also with wider aspects associated with learning how to find, use, summarize, evaluate, create, and communicate information effectively while using digital technologies. Digital literacy does not replace traditional forms of literacy; it builds upon its foundation.

The digital literacy barrier has been identified in numerous surveys attempting to explain broadband non-adoption. For example, in the United States (Horrigan 2009), 13 % of non-adopting households mentioned difficulty of use as a major barrier for adopting broadband. This answer comprised different reasons for difficulty (lack of training, age, physical handicap such as being visually impaired). Research conducted in the context of the United States National Broadband Plan found that 22 % of non-broadband adopters said that they were not comfortable using a computer because of limited digital literacy (Horrigan 2013). In Spain, digital illiteracy amounted to 29 % of the broadband non-adopting households (ONTSI 2012). This number is close to the one reported by non-adopters in Puerto Rico (31 %), which explained their behavior by saying that "they do not need broadband or the Internet" (PRBT 2012). In the United Kingdom, lack of digital ability was only cited by 4 % of non-adopting households (OFCOM 2012). This metric is close to the results of a comparable Australian survey, where 7 % of non-adopters found "the Internet to be too complicated" (AGIMO 2009). In the Colombian research cited earlier, 8.6 % of households surveyed responded that they did not know how to use computers as an explanation for not adopting broadband (MTIC 2011).

As indicated in the Fig. 2.4 above, limited digital literacy is determined primarily by two structural variables: education level and age. In addition, income level (as correlated with education level) remains a contributing factor. However, digital illiteracy could be particularly high in certain socio-demographic groups, such as the elderly, the unemployed, the disabled, and certain female groups.

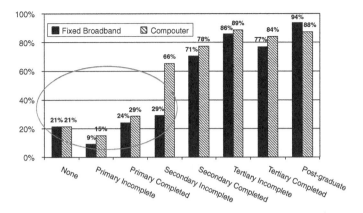

Fig. 2.11 Costa Rica: household computer and broadband adoption by level of education of head of household (2010). *Source* Costa Rica. Rectoria de Telecomunicaciones

2.2.2.1 Broadband, Digital Literacy, and Education

The first indication that educational level and broadband adoption were intrinsically linked was found when comparing country broadband demand gap and household education across countries. For example, the Republic of Korea has one of the lowest broadband demand gaps in the world: 93 % of households have adopted fixed broadband, which would indicate an adoption gap of 7 % (given the 100 % coverage level). On the other hand, Korean citizens complete on average an additional year of education compared to citizens in Japan and the United States. Additionally, Korean households have 50 % more aggregate years of education than households in the United States. This statistic would suggest a correlation at the aggregate level between education and broadband adoption.

Further research at the country level confirmed this initial evidence. For example, in Costa Rica, broadband adoption doubles when the head of household has completed high school (see Fig. 2.11).

As Fig. 2.11 indicates, while computer ownership increases substantially after the head of household completes primary school, broadband adoption jumps after some years of secondary education have been fulfilled. This fact would indicate that, beyond the general influence of the education level variable, affordability plays a stronger role (considering that schooling and income are correlated) in the case of broadband subscription than in terms of purchasing a computer. The combined impact of household income and level of education in the Costa Rican study can be clearly visualized in Fig. 2.12.

As the data in Fig. 2.12 suggest, the overall direct relationship between household income and fixed broadband adoption is clear. One exception to this trend is seen in second decile households where the head has a tertiary education: in this case, adoption is significantly higher than households below the eighth decile. This anomaly could be explained by the existence of households where the

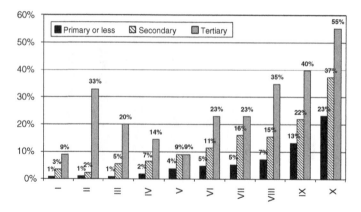

Fig. 2.12 Costa Rica: household broadband penetration by income decile and education (2010). *Source* Costa Rica. Rectoria de Telecomunicaciones

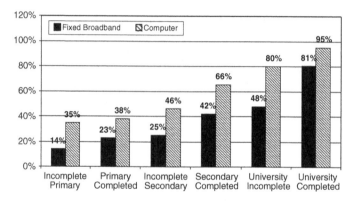

Fig. 2.13 Chile: computer and fixed broadband adoption by education level of head of household (2009). *Source* Universidad Alberto Hurtado (2009)

head is a recent university graduate who has not yet been able to earn an income commensurate to his or her educational achievement.

Beyond the direct relationship between income and broadband adoption, the influence of education is quite relevant. Particularly, above the sixth decile (where affordability represents less of a barrier), education becomes a determining factor. The higher the educational achievement of the head of household, the higher broadband adoption is.

The importance of education in explaining broadband adoption has also been detected in a study in Chile (see Fig. 2.13).

In the case of Chile, fixed broadband and computer adoption approaches the 50 % household penetration after secondary school has been completed. A similar finding was produced by a study in Puerto Rico conducted in the context of the state's broadband strategy development (Fig. 2.14).

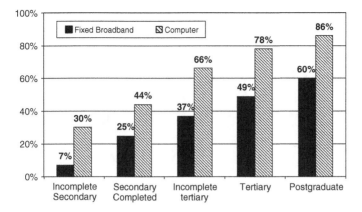

Fig. 2.14 Puerto Rico: computer and fixed broadband adoption by education level of head of household (2011). *Source* PRBT (2012)

The study of the education variable reveals the complex interrelationship it has with the affordability factor. At lower income levels, the affordability variable is stronger than the educational in predicting adoption. On the other hand, at income levels higher than the sixth decile, demand is less elastic to income, and educational achievement becomes preeminent.

Again, from a policy perspective, despite the importance of sector specific initiatives such as digital literacy programs, classical education programs will ultimately have a significant contribution to stimulating broadband demand.

2.2.2.2 Broadband, Digital Literacy, and Age

Studies conducted in the developed world have all pointed out the existence of a generational gap linked to limited digital literacy. In the United Kingdom and the United States, for example, the average age of a non-adopting household is over 65 years old (OFCOM 2012).

Research in the emerging world suggests the existence of a threshold of 30 years old, after which Internet use tends to decline significantly. For example, in Chile the percentage of non-adopters doubles after 30 years old (Universidad Alberto Hurtado 2009). In Peru, the percentage of Internet users within the 19–24-age bracket is 61 %, compared to 37 % amongst the 25–40 cohort (INEI 2012). In Brazil, the percentage of Internet adopters 24 years old or younger is 81 %, compared to 48 % amongst the 35–44 age group. In Costa Rica, the adoption of fixed broadband tends to decline significantly after 45 years old (see Fig. 2.15).

The difference between the 30-year threshold for Internet usage and persisting broadband penetration at the 35–44 bracket is explained by the presence of children in the household. Children tend to act as change agents in a household, stimulating Internet usage and sustaining broadband adoption. This indirect influence cancels some of the generational gap identified in numerous studies. In

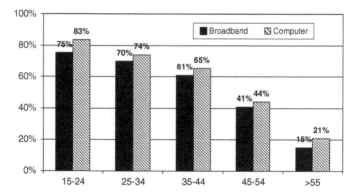

Fig. 2.15 Costa Rica: household fixed broadband and computer adoption by age cohort (2010). *Source* Costa Rica. Rectoria de telecomunicaciones (2011)

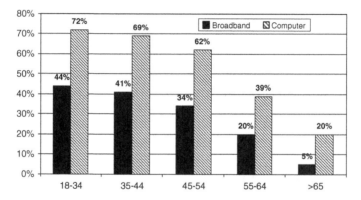

Fig. 2.16 Costa Rica: household fixed broadband and computer adoption by age cohort (2010). *Source* PRBT (2012)

the case of Chile, the presence of children in the household increases the probability of acquiring a broadband subscription from 39 to 43 % (Universidad Alberto Hurtado 2009). In Peru, the probability increases from 43 to 57 % (INEI 2012). In Costa Rica, the effect was not general across ages, but when controlling for income and education, the presence of children as a stimulus for adoption becomes very strong in households where the head has a low level of education. In Costa Rica survey data confirms this trend. Broadband adoption drops 14 percentage points after 54 years old (see Fig. 2.16).

The digital literacy generational gap poses more serious problems for some countries than others. Asian countries like Hong Kong, Japan, and Singapore, for instance, have some of the world's fastest growing aging populations. Even in these countries, seniors appear resistant to broadband use and adoption. Research suggests that governments and institutions could potentially change this trend by focusing on the socialization of technology to make it more widespread throughout this demographic (Computer Supported Cooperative Work 2011).[7] Programs addressing the generational gap will be presented in Sect. 3.1.2.

2.2.3 Lack of Content Relevance or Interest

Since broadband is a platform used to access Internet content, applications, and services, the relevance of such content offers an incentive to purchase a subscription. Conversely, the lack of cultural relevance could serve as a barrier to adoption. Cultural relevance could be conceptualized either in terms of content suited to the interests of the adopting population or in terms of language used for interacting with applications/services or consuming content.

The relevance dimension has been identified in several studies in the developed world. For example, Horrigan (2009) estimated that, according to his survey, 50 % of non-broadband households linked non-adoption to "lack of relevance/interest." In the survey, lack of relevance was driven by "no interest", "busy conducting other tasks", or other unspecified reasons. Interestingly enough, the percentage of non-adopting households citing lack of relevance (50 %) was higher than the percentage citing affordability (35 %). In a study conducted in 2011, the non-broadband adopting households that provided the "lack of relevance" explanation only decreased to 47 %, while affordability dropped to 24 % (NTIA 2011). When disaggregating non-broadband households between those that have or do not own a computer, "lack of relevance" jumps to 52 %, and affordability drops to 21 %.

"Lack of relevance" consistently outranks affordability structural factors in most studies conducted in developed nations. For example, in a study conducted in the United Kingdom in 2011 (OFCOM 2012), 66 % of non-adopting households said "lack of relevance" explained their decision not to purchase a broadband service. Again, this percentage was substantially less than those households that alluded to the affordability barrier (16 %). Interestingly enough, in a similar study conducted in 2010 (OFCOM 2012), the affordability barrier was mentioned by 23 % of households surveyed. This suggests that, as prices for broadband service decline, the cultural relevance factor gains in importance. In other words, from a policy standpoint, once the economic obstacles are tackled and affordability becomes less of an explanatory factor of non-adoption, the lack of relevance or interest variable gains weight. In the case of Australia, affordability (26 %) was somewhat more important than lack of relevance (19 %) (AGIMO 2009).

As expected, lack of relevance as a barrier in developed countries is prevalent in very circumscribed socio-demographic categories. In the United Kingdom, the non-broadband households that cite lack of relevance tend to be lower income households with people over 65 years old. In a study conducted in Spain

---

[7] https://sites.google.com/site/technologyamongseniors/

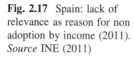

Fig. 2.17 Spain: lack of relevance as reason for non adoption by income (2011). *Source* INE (2011)

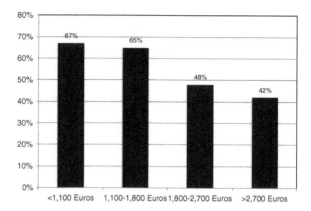

(ONTSI 2012), lack of relevance of Internet content is inversely proportional to income levels (see Fig. 2.17).

In at least one study conducted in a developed country, the linguistic factor contributed to the lack of relevance. That was identified in the United States among the Hispanic population that had recently immigrated to the country. It is important to consider, however, that, as in the United States, the linguistic barrier is strongly correlated with economic and educational factors. Therefore, it is still difficult to tease out the socio-demographic variables in order to isolate the linguistic factors.

The language barrier has been identified in the emerging world as well. For example, in Peru, only 8 % of those individuals whose first language was not Spanish are Internet users. That percentage increased to 40 % among native Spanish-speakers. In the Middle East North Africa (MENA) region, the relatively low availability of native-language content is cited as a major challenge to broadband demand (World Bank, Broadband Strategies Toolkit Chap. 7). Focusing on the development of local-language digital content may be the key to increasing the uptake of broadband in these instances. This issue will be discussed at length in Chap. 6.

Beyond language, lack of content relevance remains a strong variable influencing non-adoption. For example, in Colombia, 20 % of non-broadband households justified their behavior explaining that they did not view the Internet "as being necessary" (MITIC 2011).

The lack of relevance variable presents some complexity in terms of its understanding. Two interpretation options are open to policy makers. One option is that the consumer has evaluated the offerings in terms of applications, services, and content and has not found them relevant to his/her needs. Under this premise, policy initiatives should be oriented toward increasing the perceived value of broadband by expanding the range and utility of offerings (these are called "demand pull" policies). The second option is that the consumer does not have enough information to make a decision of adopting broadband. The policy implication in this case is that the consumer needs to be made aware of the potential of the technology (called "awareness" policies).

2.2.4 Broadband Diffusion Cycles and the Importance of Adoption Structural Factors

The importance and weight of each of the three residential demand structural factors—affordability, awareness, and relevance—is not homogeneous across the broadband diffusion cycle. Some are more important than others, depending on the level of adoption of the technology in a given country. This is a critical concept that needs to be understood before deploying demand stimulation policies.

The studies of both fixed and wireless broadband adoption in developed countries would indicate that residential broadband adoption tends to proceed along three clearly defined stages (see Table 2.9).

In stage 1, at lower levels of adoption, the factor constraining penetration is supply-driven. Price does not play a significant role because the first group of adopters is relatively price insensitive. In his research on broadband adoption, Varian (2002) found that the first group of subscribers is fairly price insensitive, while the next generation exhibits an elasticity of demand between -1.3 and -3.1. The second variable affecting broadband adoption is device availability. For example, Chinn and Fairlie (2010) found in their study of Internet usage in 161 countries between 1999 and 2001 that the main factors affecting adoption are possession of a computer and awareness of benefits. Ono and Zavodny (2007) confirmed this finding; relying on microdata for United States from 1997 to 2003, they found that the possession of a computer to use Internet remains the main barrier. Vicente and Lopez (2006) obtained a similar result for Europe using data for 15 countries in 2002. Obviously, current research should indicate that smartphone ownership becomes the first barrier to broadband adoption.

In stage 2, beyond a coverage tipping point, the most important, variable driving penetration is affordability. When Chinn and Fairlie (2010) extended their study up to 2004, they found that the price of the service started to be relevant in the explanation of the levels of Internet adoption. When affordability becomes a more important barrier, elasticity coefficients increase dramatically. For example, using survey data from 100,000 households in the United States, Goolsbee (2002) found that in areas where service is available to a majority of households, a decline in broadband prices of 10 % yields an increase in penetration ranging between 21.50 and 37.60 % (with a mean value of 26.50 %). A similar price reduction in areas where the service is not available to a majority of consumers would result in an increase in penetration ranging between 15 and 30 %. There are three reasons why elasticity is higher in areas with full service coverage. First, in areas with partial coverage, early adopters are less sensitive to prices and therefore, demand is inelastic. Second, in areas with full coverage, consumers have the opportunity of observing the benefit broadband generates, and they are willing to engage in cost/benefit analysis, whereby any reduction in pricing would increase the consumer surplus. Third, in areas with full broadband coverage, consumers also consider network effects when conducting a cost/benefit analysis.

Table 2.9 Stages of broadband adoption

	Stage 1	Stage 2	Stage 3
Ownership of access devices (computers, smartphones)	Low adoption	Medium adoption	High adoption
Availability of web applications and services	• Very low	• Limited	• High
Factors driving non-adoption	• Service coverage	• Affordability	• Digital literacy • Cultural relevance

Source Developed by the authors

Rappoport et al. (2002) also confirm the importance of the affordability barrier in Stage 2. In a survey of 5,225 urban households in the United States, the authors determined that a 10 % price reduction of broadband would yield an increment of 14.91 % in service adoption. Extending the analysis to the OECD countries between 2003 and 2008, Lee et al. (2011) found that 10 % price reduction of broadband results in a 15.80 % increase in penetration. In this last research, the impact of price reduction begins to diminish relative to earlier studies, anticipating the transition to next stage.

In Stage 3, at higher penetration levels, price elasticity coefficients start to diminish. In their study of price elasticity in the United States between 2005 and 2008, Dutz et al. (2009) observed that coefficients declined from -1.53 in 2005 to -0.69 in 2008. Coincidently, Cadman and Dineen (2008) estimated elasticity coefficients for OECD countries in 2007 to be -0.43. This could be due to a shift in consumer perception as to the value of broadband (and consequently the subscriber willingness to pay) from "luxury" to "necessity". On the other hand, because of diminishing importance in the affordability barrier to adoption, as reviewed in the survey data earlier, structural factors related to limited digital literacy and cultural relevance take precedence.

Consequently, according to the research conducted up to now, the evidence regarding broadband adoption structural factors could be conceptualized as follows (see Fig. 2.18).

In Stage 1, the primary lever to foster adoption is service coverage. In Stage 2, affordability becomes the most important barrier, although digital literacy and cultural relevance begin to assume greater prevalence. In Stage 3, at higher penetration levels, price sensitivity becomes secondary, and the most important adoption barriers remain digital literacy and cultural relevance.

Moving to emerging countries, initial evidence produced by Galperin and Ruzzier (2011) confirms that regions whose penetration is within stage 2 (3–20 %) exhibit high elasticity. OECD countries, with an average penetration of 27.48 % have a price elasticity of -0.53, while Latin American countries (average penetration of 7.66 % in 2011) have an elasticity of -1.88.

While there is still not evidence available, two factors could change the sequential pattern of adoption structural factors outlined earlier. First, as a result of the increasing adoption of mobile broadband enabled devices (such as

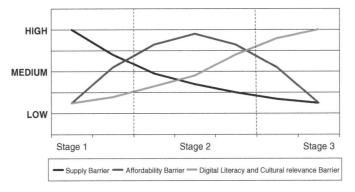

Fig. 2.18 Relative importance of broadband adoption structural factors in reaching advanced stages of broadband penetration. *Source* Developed by the authors

smartphones) and the deployment of 3G and 4G networks, the structural factors of Stage 1 could shorten up significantly, rapidly putting countries in the need to tackle the affordability barrier. Secondly, the increasing availability of applications and services could enhance the willingness to pay of subscribers, thereby altering the consumer surplus equation and reducing the elasticity coefficients. This could be an important factor in many emerging countries.

A final note should be made that, while the generic evolution model presented earlier is one followed by a large number of countries, in some contexts residential broadband adoption may skip Stage 1 and start directly from Stage 2 or skip Stage 1 and 2 and move directly to Stage 3. For example, in Japan, when DSL service started from scratch in areas only with dial-up connection, some ISPs experienced a jump-start in penetration rate of between 10 and 20 %.

2.3   Understanding the Enterprise Broadband Demand Gap

The structural factors and adoption obstacles of broadband among enterprises are different than those variables constraining diffusion among individual consumers. Notwithstanding the fact that broadband technology is a production factor with a positive contribution to the efficiency of business operations, small and medium businesses (especially microenterprises) in emerging economies have faced some broadband adoption impediments. Kotelnikov (2007) has defined four stages of ICT adoption within the small and medium business universe. Those stages are depicted in Table 2.10.

These four stages should not be considered in a static fashion, particularly in light of technological progress and price reductions. Nevertheless, data would support the notion that, particularly in emerging markets, these four stages remain a fairly common development path. This is illustrated by the still limited adoption

Table 2.10 Stages of ICT adoption in SMEs

	Basic communications	Basic information technology	Advanced communications	Advanced information technology
Telecommunications	• Wireline • Wireless • Facsimile		• Email • Broadband • Videoconferencing • File sharing • E-Commerce • VoIP	
Information technology		• Personal computer • Basic software (spreadsheet, word processing)		• Data base management • ERP • nventory management • CRM

Source Kotelnikov (2007)

of broadband within SMEs. Furthermore, while in many cases small businesses rely on the Internet, they do so in shared facilities as opposed to on their own business premises (see Table 2.11).

At the outset, the nature of the business of SMEs, especially micro-enterprises, explains the lack of broadband adoption. Katz (2009) argues that SMEs can be grouped in three categories, each of which has a different need for broadband services:

- "International" SMEs: regardless of their size, these units need broadband to gain access to international markets, deal with their supply chain, and support their logistics
- SMEs supplying large enterprises: as part of the supply chain of large firms, these SMEs require broadband to receive orders, process payments, and provide delivery information[8]
- SMEs operating in low value added industry sectors, primarily services: these firms have a low compelling need to purchase broadband since the nature of their business does not require the type of transactions mentioned in the other two categories.

Considering that the last category comprises a large portion of the SME population in emerging economies, it is natural to expect a delay in the adoption of broadband among small enterprises.

However, beyond this structural factor, other reasons explain the low level of broadband adoption, including limited access to investment capital, comparatively high technology costs, and lack of training. Regarding capital investment and monthly service costs, it is important to note that a significant proportion of SMEs

---

[8] A particular case of SMEs refers to start-ups initiated by an incubator program. See below in Sect. 3.4.3.

Table 2.11 ICT adoption among SMEs

Country	Personal computers (%)	Internet (%)	Broadband (%)	Year
Argentina	43	97	75	2007
Brazil	69	54	9	2009
Canada	99	99	99	2010
Chile	74	66	60	2006
Colombia	16	…	9	2010
Ecuador	…	47	…	2005
El Salvador	47	36	50	2005
Guatemala	32	15	16	2005
Mexico	87	73	45	2005
Nicaragua	39	15	11	2005
Peru	27	23	60	2009
Puerto Rico	…	…	74	2011
United States	…	…	75	2010
Venezuela	5	12	3	
Average	46	48	30	

Sources
Argentina Indec; Prince and Cook
Brasil SEBRAE
Canada Fleet (2012)
Chile http://www.google.com/url?sa=t&rct=j&q=&esrc=s&source=web&cd=1&cad=rja&ved=
0CC8QFjAA&url=http%3A%2F%2Fwww.cnc.cl%2FCharla%2520PYMES%2520Mayo-08.
ppt&ei=waiiUI77GY-88wTP84CgDQ&usg=AFQjCNGa00NabrVh3VzGOF2FSp5421_efg&
sig2=Sxn1GM5GmynCRvKAS1J0sA
Colombia National Department of Statistics
Ecuador FENAPI
El Salvador, Guatemala, Nicaragua Monge-Gonzelez et al. (2005)
Mexico Select
Peru http://gestion.pe/noticia/304158/conectividad-pymes-banda-ancha-creceria-10-este-ano
Puerto Rico Puerto Rico Broadband Task Force (2012)
United States Connected Nation (2010) *Puerto Rico*: Puerto Rico Broadband Task Force (2012)
Venezuela Microsoft
Note Due to the fact that the sources and methodologies for estimating these statistics are not
consistent, data in this chart should not be compared across countries

do not receive fixed monthly income because they operate outside of the formal economy of emerging countries. Their income is generally daily or weekly and is dependent upon the type of labor performed; thus, they cannot borrow long-term or purchase products that require a fixed monthly payment such as PCs, servers or Internet access. These enterprises are generally forced to use prepaid wireless, Internet booths, or cybercafés, and rented PCs.

Secondly, many of the entrepreneurs that run SMEs (which are primarily microenterprises in emerging countries) have very limited level of technological training. A large number of SME owners in these economies face a generational gap by not receiving Internet technology exposure growing up. Therefore, they lack the necessary training to operate a computer or use broadband to improve business efficiency. This lack of education translates into the anxiety of using technology and ignorance of its capability to create economic value. The limited availability and retention of a skilled ICT workforce also pose a problem for SMEs, particularly in emerging markets. Because of the systemic shortage of technical personnel, large companies offer wages to graduates of higher education that SMEs cannot match. Even when SMEs manage to hire graduates, retention rates are very low.

Broadband adoption by SMEs is also limited by the lag required to make the necessary organizational and business process changes to assimilate broadband and data transmission technologies. In general terms, SMEs (particularly in emerging countries) tend to restrict the use of ICT to accounting and finance, while neglecting its application to production processes. A survey by the Chilean Ministry of Economy found that only 2.6 % of Chilean companies used ICT to increase the efficiency of business processes other than accounting and finance. Yet, the survey made an even more worrisome observation: 80 % of companies reported that they did not implement ICT in areas other than finance and accounting because they lacked the technological expertise necessary to understand its benefits.

In sum, beyond the composition of the SME sector, which might structurally constrain the need to adopt broadband, the enterprise broadband demand gap is the result of three obstacles:

- Limited affordability: certain portions of the SME space either cannot acquire a device or purchase the subscription needed to access the Internet
- Limited technology training constrains the ability to purchase and effectively introduce broadband in the firm
- The assimilation of broadband to render efficiency gains in the small business requires the introduction of changes in organization, business processes, and even use of IT, all tasks that are well beyond the scope of expertise of small business management

These relationships have been depicted in Fig. 2.19.

Recognizing the benefits broadband holds for SMEs will have additional positive macro-level effects on the country beyond just penetration rates. Indian SMEs, for instance, spent a combined US$ 9.9 billion on the IT sector in 2009 alone, and this spending was attributed to the increased demand for high speed Internet and broadband (Access Markets International 2010).[9] Many of the firms also invested in hiring more employees to utilize this technology and industries such as e-commerce boomed.

The following section will explain each of the three SMEs obstacles—affordability, training, and assimilation. Each section presents studies and data regarding the obstacles and driving variables in developed and emerging countries alike.

---

[9] http://news.indiamart.com/story/india-smbs-recover-economic-downturn-and-move-ahead-12259.html

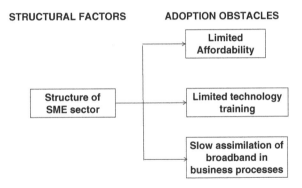

Fig. 2.19 Broadband adoption structural factors within enterprises

2.3.1 The Economic Barrier

As mentioned earlier, the affordability barrier is not only important for individual consumers, but also relevant in the case of SMEs, although this variable tends to be more important in the case of microenterprises.

In a survey conducted in Colombia among SMEs, 12.9 % of surveyed firms indicated that they did not have the economic means to pay for broadband service, while 9.3 % would like to buy broadband because of its utility, but found the service pricing to be too high. While the research does not break down the answers by size of firm, broadband adoption numbers help determine that most of the firms mentioning affordability as a barrier are concentrated among microenterprises: broadband penetration among firms with 11–50 employees is 79 %, while adoption among firms with less than 11 employees is only 7 % (National Department of Statistics 2010).

2.3.2 Limited Technology Training

The difficulties encountered in recruiting staff with technical skills to select, purchase, and operate ICT infrastructure serves as a critical limitation for adopting broadband. This factor was measured in a survey regarding the difficulty to recruit ICT trained personnel conducted among SME managers in Latin America (see Fig. 2.20).

As Fig. 2.20 indicates, the recruiting constraint is particularly acute in Argentina and Brazil. Katz (2009) also identified the problem in field research conducted in other countries such as Uruguay and Chile. The constraint in recruiting technical personnel is due to the fact that the educational system does not generate enough graduates in ICT-related disciplines. In that context, salary inflation "prices out" SMEs when it comes to attracting graduates, which end up working for large enterprises.

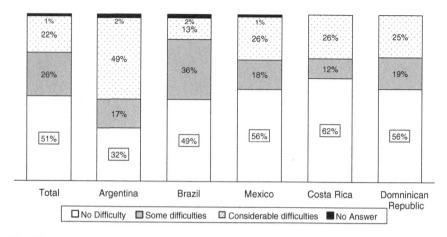

Fig. 2.20 Latin America: difficulty to recruit ICT trained personnel. *Source* UPS Business Monitor (2007)

In developing countries, the problem is not exclusive to smaller businesses. South Africa faces a similar situation. In 2012, the country reported an unemployment rate of 25 %, with more than 4.2 million adult citizens actively searching for employment to no avail. While the government attempted to address the situation by "creating jobs," a more in-depth analysis demonstrated that a lack of qualified workers contributed more to the problem than did a lack of available jobs. In fact, the country's business organizations reported more than 800,000 vacant positions, the majority of which were found in high-skilled industries such as information technology, engineering, and finance.

A 2012 survey of Malaysian CIOs concluded that the low digital literacy skills of organization executives negatively impacted business capabilities (Vanson Bourne 2012).[10] The CIOs "feared senior-level digital illiteracy is causing a lack of market responsiveness, missed business and investment opportunities, poor competitiveness and slower time to market." This sentiment was felt across many Asian markets, where business leaders appeared to fall behind their peers in more developed economies.

2.3.3 Slow Assimilation of Broadband

In order to increase efficiency and output, the adoption of information and communication technologies by enterprises requires the introduction of a number of processes and organizational changes. These changes, as well as training and other

---

[10] http://www.cio-asia.com/resource/management-and-careers/leaders-low-digital-literacy-may-hamper-business-growth-malaysian-study/

cultural factors (such as entrepreneurial spirit, willingness to take risks in an organizational transformation), are referred to as the "accumulation of intangible capital."[11] Broadband alone does not have an economic impact. It rather enables the adoption of e-business processes that result in increased efficiency (such as streamlined access to raw materials and management of the supply chain, or better market access). Intangible capital accumulation and the adoption of e-business processes delay the full economic impact of broadband.

This gradual process of technology adoption and assimilation can be studied in the aggregate for economies as a whole. Certain companies, by virtue of the innovativeness of their management and their willingness to transform their enterprises, are the leaders that will initially reap the benefits of ICT.

The second wave of adoption is concentrated on industrial sectors whose structure and value chains tend to result in higher transaction costs. These network-oriented industries are concentrated in financial services, transportation, or retail distribution sectors. In these industries, complexity costs are so high that, in addition to increasing the number of information workers, they need to adopt technology to improve their productivity. This wave represents a move from firm-related adoption drivers to industry structure and economics. It is only in those economies that Jorgenson et al. (2007) call "IT intensive", where the concentration of industrial sectors more prone to adopt ICT is higher, that we can see the macro-impact of ICT on productivity.

Small and medium enterprises tend to adopt broadband in the third wave, after they have been able to make the necessary process and organizational changes needed to assimilate broadband-enabled applications. In light of this effect, these firms will naturally lag in the assimilation of broadband technology. The public policy implications of this effect cannot be understated. To achieve full economic benefit of broadband deployment, governments need to emphasize the implementation of training programs and, in the case of SMEs, offer consulting services that help firms capture the full benefit of the technology. Details of these specific proposals will be reviewed in Chap. 3.

2.4 Broadband Demand Gap Diagnostic

A diagnostic of broadband demand obstacles needs to precede the formulation of suitable policies. Building on the concepts presented earlier, the development of broadband demand stimulation policies needs to begin by conducting a diagnostic of the demand gap. This should follow a structured methodology tackling the following questions:

---

[11] Basu and Fernald (2007).

- Quantification of the residential demand gap

  - What is the coverage of fixed and mobile broadband services (portion of the population served or unserved)?
  - What is the percentage of the population served by both fixed and mobile platforms?
  - What portion of the population by administrative unit (such as departments or counties) purchases both fixed and mobile services (complementarity effect)?
  - What portion of the population purchases only mobile broadband services (PC connectivity devices, smartphones) (substitution effect)?
  - What portion of the population is served by fixed broadband and do not acquire a subscription?
  - What portion of the population is served by mobile broadband exclusively and do not acquire a broadband plan?

- Socio-demographic analysis of the residential demand gap

  - What is the profile of non adopters (by age, income level, education, gender, ethnic group)
  - Disaggregation of non-adopting population by region and administrative unit

- Quantification of the enterprise demand gap

  - What is the percentage of enterprises (primarily SMEs served by both fixed and mobile platforms?
  - What portion of the enterprises by administrative unit (such as departments or counties) purchases both fixed and mobile services (complementarity effect)?
  - What portion of the enterprises purchases only mobile broadband services (PC connectivity devices, smartphones)?
  - What portion of the enterprises is served by fixed broadband and do not acquire a subscription?
  - What portion of the enterprises is served by mobile broadband exclusively and does not acquire a broadband plan?

- Firm level analysis of the residential demand gap

  - What is the profile of non adopting enterprises (by size, and industrial sector)
  - Disaggregation of non-adopting enterprises by region

By answering these questions, policy makers will be able to develop a diagnostic of the broadband demand gap. With this diagnostic, policy makers need to start devising appropriate demand stimulation policies. The different examples of policy initiatives are reviewed in the following chapters.

References

Australian Government Information Management Office. 2009. *Interacting with Government.* [report]

Access Markets International: http://computer.financialexpress.com/20100614/news06.shtml (2010)

Basu, S., Fernald, J.: Information and communications technology as a general-purpose technology: evidence from US industry data. Ger. Econ. Rev. **8**(2), 146–173 (2007)

Bohlin, E., Srinuan, P., Srinuan, C.: The mobile broadband and fixed broadband battle in Swedish market: complementary or substitution? http://fsr.eui.eu/Documents/WorkingPapers/ComsnMedia/2011/WP201136.pdf. Accessed 22 March 2014

Cadman, R., Dineen, C.: Price and income elasticity of demand for broadband subscriptions: a cross-sectional model of OECD countries. http://spcnetwork.eu/uploads/Broadband_Elasticity_Paper_2008.pdf (2008). Accessed 22 March 2014

CGI. 2012. Survey on the Use of Information and Communication Technologies in Brazil, CGI.br

Chaudhuri, A., Flamm, K.S., Horrigan, J.: An analysis of the determinants of Internet access. Telecommun. Policy **29**(9), 731–755 (2005)

Chinn, M.D., Fairlie, R.W.: ICT use in the developing world: an analysis of differences in computer and Internet penetration. Rev. Int. Econ. **18**(1), 153–167 (2010)

Computer Supported Cooperative Work: Technology among seniors in asia. https://sites.google.com/site/technologyamongseniors/ (2011). Accessed 22 March 2014

Dutz, M., Orszag, J., Willig, R.: The Substantial Consumer Benefits of Broadband Connectivity for US Households. Internet Innovation Alliance, New York (2009)

Euromonitor International: Passport: Market Research Database (2011)

Firstbiz. Indian SMEs spent $9.9 billion on IT in 2009: AMI Partners. http://www.firstbiz.com/biztech/indian-smes-spent-9-9-billion-on-it-in-2009-ami-partners-9852.html (2010). Accessed 22 March 2014

Fleet, G.: Growth of broadband and ICT adoption by SMEs in Atlantic Canada. J. Knowl. Manag. Econ. Inf. Technol. **2**(3):1–9 (2012)

Galperin, H., Ruzzier, C.A.: Broadband tariffs in Latin America: benchmarking and analysis. http://papers.ssrn.com/sol3/papers.cfm?abstract_id=1832737 (2011). Accessed 22 March 2014

Goolsbee, A. Subsidies, the value of broadband, and the importance of fixed costs. In: Robert, C., James, H.A. (eds.) Broadband: Should We Regulate High-Speed Internet Access, pp. 278–294. Brooking Institution Press, Washington (2002)

Grazzi, M., Vergara, S.: Determinants of ICT access. In: Balboni, M., Rovira, S., Vergara, S. (eds.) ICT in Latin America: A microdata analysis. ECLAC, Santiago (2011) http://www.cepal.org/publicaciones/xml/7/43847/R.2172ICTinLA.pdf Accessed 20 March 2014

Hauge, J.A., Prieger, J.E.: Demand-side programs to stimulate adoption of broadband: what works? Rev. Netw. Econ. **9**(3) (2010)

Hilbert, M.: Digital gender divide or technologically empowered women in developing countries? A typical case of lies, damned lies, and statistics **34**(6), 479–489 (2011)

Horrigan, J.: Broadband adoption in Illinois: Who is online, who is not, and how to expand home high-speed adoption. [report] Partnership for Connected Illinois (2012)

Horrigan, J. Broadband adoption and use in America. OBI Working Paper Series No. 1. http://online.wsj.com/public/resources/documents/FCCSurvey.pdf (2009). Accessed 22 March 2014

Horrigan, J.: Adoption of information and communication service in the United States: Narrowing Gaps, New Challenges. [online] Knight Foundation, August. http://knightfoundation.org/media/uploads/media_pdfs/DigitalAccessUpdateFeb2014.pdf (2013). Accessed 1 Apr 2014

INEGI. Encuesta Nacional de Ocupación y Empleo (ENOE). http://www.inegi.org.mx/est/contenidos/Proyectos/encuestas/hogares/regulares/enoe/ (2011). Accessed 22 March 4

INEI (Instituto Nacional de Estadística e Informática): Las tecnologías de información y comunicación en los hogares. Lima, Peru (2012)

Jorgenson, D., Ho, M., Stiroh, K.: Industry origins of the American productivity resurgence. Econ. Syst. Res. **19**(3):229–252 (2007)

Katz, R.L.: La Contribución de las tecnologías de la información y las comunicaciones al desarrollo económico: propuestas de América Latina a los retos económicos actuales. Ariel, Madrid, España (2009)

Katz, R., Callorda, F.: Impacto del despliegue de la banda ancha en el Ecuador. Lima: Dialogo Regional sobre Sociedad de la Informacion (2013)

Katz, R.L., Galperin, H.: Addressing the broadband demand gap: drivers and public policies. http://papers.ssrn.com/sol3/papers.cfm?abstract_id=2194512 (2012). Accessed 20 March 2014

Kotelnikov, V.: Small and medium enterprises and ICT. e-Primers for the information economy, society and polity. [report] Asia-Pacific Development Information Programme (2007)

Lee, S., Marcu, M., Lee, S.: An empirical analysis of fixed and mobile broadband diffusion. Inf. Econ. Policy, **23**(3):227–233 (2011)

MITIC: (Ministerio de Tecnologías de la Información y la Comunicación). Vive Digital: Documento Vivo del Plan versión 1.0 [report] Bogota, Colombia (2011)

Monge-Gonzalez, R., Alfaro-Azfeifa, C., Alfaro-Chamberlain, J.: The Central American SMEs and ICTs: An Empirical Study on the Impact of ICTs Adoption on SMEs Performance. International Development Research Center, San Jose Costa Rica (2005)

National Department of Statistics: (Departamento Administrativo Nacional de Estadística, Colombia). Enterprise adoption of ICT. http://www.dane.gov.co/index.php/tecnologia-e-innovacion-alias/tecnologias-de-la-informacion-y-las-comunicaciones-tic (2010). Accessed 26 March 2014

Navarro, L., Sanchez, M.: Gender differences in Internet use. In: Balboni, M., Rovira, S., Vergara, S. (eds.) ICT in Latin America: A microdata Analysis. Santiago: ECLAC. http://www.cepal.org/publicaciones/xml/7/43847/R.2172ICTinLA.pdf (2011). Accessed 20 March 2014

NTIA: Exploring the digital nation. (report). Washington, DC (2011)

OFCOM: Adults media use and attitudes report. http://stakeholders.ofcom.org.uk/binaries/research/media-literacy/media-use-attitudes/adults-media-use-2.pdf (2012). Accessed 22 March 2014

Ono, H., Zavodny, M.: Digital inequality: a five country comparison using microdata. Soc. Sci. Res. **36**(3), 1135–1155 (2007)

ONTSI: (Observatorio Nacional de las Telecomunicaciones y de la Sociedad de la información). La Sociedad en Red. Informe Anual 2011. (report) Madrid (2012)

PRBT: Puerto rico broadband strategic plan. http://www.connectpr.org/sites/default/files/connected-nation/Puerto%20Rico/files/chapter_3.pdf (2012). Accessed 22 March 2014

Rappoport, P., Kridel, D.J., Taylor, L.D., Alleman, J., Duffy-Deno, K.T.: Residential demand for access to the Internet. Int. Handb. Telecommun. Econ. **2**, 55–72 (2002)

Subtel.: Encuesta sobre Acceso, Uso y Usuarios de Internet Banda Ancha en Chile. Realizado por Universidad Alberto Hurtado. Junio (2009)

Rectoría de Telecomunicaciones: II Evaluación de la Brecha Digital en el Uso de Servicios de Telecomunicaciones de Costa Rica. San José, Costa Rica. febrero (2011)

Universidad Alberto Hurtado/Subtel: (Department of Telecommunications). . Encuesta sobre acceso, uso y usuarios de internet banda ancha en Chile. Santiago, Chile (2009)

Varian, H.: The net impact study. Cisco Systems Inc, San Jose (2002)

Vanson Bourne: Leaders' low digital literacy may hamper business growth: Malaysian study. http://www.cio-asia.com/resource/management-and-careers/leaders-low-digital-literacy-may-hamper-business-growth-malaysian-study/ (2012). Accessed 7 June 2014

Vicente, M.R., Lopez, A.J.: Patterns of ICT diffusion across the European union. Econ. Lett. **93**(1), 45–51 (2006)

Williams, M.D.J.: Broadband for Africa: developing backbone communications networks. https://openknowledge.worldbank.org/bitstream/handle/10986/2422/536430PUB0Broa101Official0Use0Only1.pdf?sequence=1 (2010). Accessed 22 March 2014

Chapter 3
Creating Awareness

In Chap. 2 the structural factors constraining broadband adoption were reviewed for both the residential and enterprise markets. This section first focuses on a specific adoption obstacle for residential subscribers: limited digital literacy. Addressing this obstacle requires the implementation of programs that build an understanding of the service offerings. However, developing awareness also requires developing user confidence, explaining the benefits of use, and promoting an understanding of security and privacy constraints. As a result, four types of initiatives, targeting digital literacy impediments will be reviewed (see Fig. 3.1).

- Digital literacy through education programs entail the inclusion of specific programs at all levels of the formal education system, requiring also the implementation of training programs for teachers
- Targeted digital literacy interventions comprise the implementation of programs addressed to specific segments of the population, such as the elderly, the economically disadvantaged or the rural population
- Deployment of community access centers allows supplying non-adopting population with devices and access points to the Internet; in addition, the access centers can become points of delivery of training programs and user support
- The privacy and security training programs allow building the levels of trust from consumers in order to foster adoption of broadband

As mentioned in Chap. 2, beyond digital literacy programs focused on residential subscribers, building awareness has also an enterprise focus, primarily targeting small and medium enterprises (SMEs). In this case, the awareness emphasis comprises initiatives in training and the promotion of broadband assimilation (see Fig. 3.2).

- Advanced ICT training is aimed at supplementing the formal education system with training of technical personnel that will facilitate the introduction and assimilation of broadband-enabled applications in small and medium enterprises
- In addition to generic advanced programs, digital training for SMEs is specifically targeted for management and personnel working in those firms
- Consulting services provided to SMEs allow for those firms to deploy and efficiently integrate broadband-enabled applications in their businesses

R. L. Katz and T. A. Berry, *Driving Demand for Broadband Networks and Services*,
Signals and Communication Technology, DOI: 10.1007/978-3-319-07197-8_3,
© Springer International Publishing Switzerland 2014

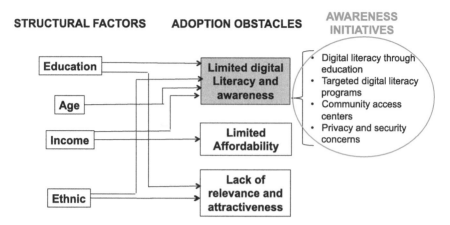

Fig. 3.1 Awareness promotion initiatives in residential broadband

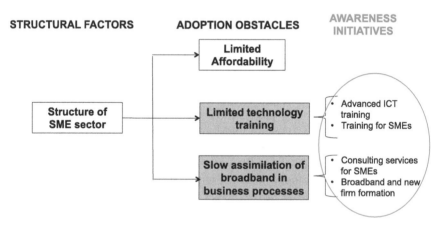

Fig. 3.2 Awareness promotion initiatives in enterprise broadband

- As a specific case of broadband impact, some initiatives should focus on how to build broadband-enabled businesses, thereby stimulating new firm formation

This section will explore all these potential approaches oriented toward raising awareness of broadband services. In doing so, cases and practices from both developed and emerging countries will be reviewed and assessed.

3.1 Developing Basic Digital Literacy

In Sect. 2.2.2 it was explained that up to 29 % of broadband non-adopters in certain countries cited limited digital literacy as a reason from not acquiring service. Again, digital literacy is defined as the "ability to use digital technology, communication tools or networks to locate, evaluate, use and create information" (Hauge and Prier 2010). Evidence from Sect. 2.2.2 also showed that, when the affordability barrier is lowered through price reductions or state fostered policies, digital literacy remains the dominant impediment. Finally, research also shows that limited digital literacy is intrinsically linked to level of education, age, and ethnic affiliation.

In light of this evidence, initiatives aimed at building digital literacy need to involve both embedding programs in the formal education system and targeting non-formal initiatives to specific segments of the population (elderly, handicapped, rural poor, etc.). The structuring of digital literacy efforts should be conducted after concluding the basic diagnostic of demand gap.

3.1.1 Digital Literacy Through Formal Education Programs

Programs oriented to fostering digital literacy through formal education consist in embedding ICT training in curricula at the primary and secondary school level complemented with targeted programs focused on teachers. This section addresses the need to introduce fundamental changes in the formal educational system in order to enhance the level of digital literacy.

Digital literacy programs embedded in the formal educational system should be, by definition, large scale and centrally driven, generally hosted within ministries of education. While providing access infrastructure (both devices and broadband), programs tend to generally focus on improving usability. As expected, the initiatives are less focused on delivering standard computer courses, emphasizing the use of IT and broadband access within course material by leveraging e-learning platforms and social networking.

3.1.1.1 Primary School Programs

Primary school digital literacy programs are critically important in building broadband awareness for numerous reasons. In the first place, a large portion of the population in many emerging nations benefit from only a primary school education (see Table 3.1).

As such, primary schooling represents the only opportunity for large segments of the population to get access to digital literacy training. Moreover, research has shown that students that gain access to broadband in school are more likely to use it later in their life (Goldfarb 2006). This appeared to be particularly true among low-income households.

Secondly, and as a corollary from the first point (and shown in Sect. 2.2.2), a large portion of the broadband non-adopting population has only a primary school education. In that sense, limited digital literacy and low levels of educational attainment are causally linked. Therefore, a digital literacy program focused on primary schools would help lowering the educational barrier.

Table 3.1 School enrollment by region of the World (2010) (Percent of population)

	North America	Latin America and Caribbean	MENA	Eastern Europe and Central Asia	European Union	East Asia and Pacific	World
Primary (%)	100.2	113.9	104.4	102.3	104.1	110.3	106.0
Secondary (%)	98.7	89.6	77.1	97.1	104.6	80.3	70.4
Tertiary (%)	67.7	40.6	30.6	58.3	61.4	29.0	29.2

Source World Bank http://data.worldbank.org/indicator/SE.SEC.ENRR?display=graph
Note These numbers can exceed 100 because they include students who fall above or below the typical age range for that level of education

Thirdly, as shown in Sect. 2.2.2, children tend to act as change agents in a household, bringing all the positive influence that stimulates Internet usage and sustains broadband adoption. By introducing intensive digital literacy programs in primary education, the initiative consists in training residential change agents that will promote literacy within low-income households. Belo and Ferreira (2012) found in researching the impact of broadband in schools in Portugal that broadband use in schools leads to higher levels of adoption in the surrounding region, and that the spillover effect is mediated by children. According to the authors, school broadband use increase the probability of adopting high-speed Internet access by 20 % in households with children. This translates into an increase of 5 % in the penetration of residential broadband within the whole country. Spillover effects have also been pointed at by Goolsbee and Klenow (2002), while the positive influence of children on residential broadband adoption was also identified in research by OFCOM (2012). Similarly, Horrigan (2014) found that children and teachers are highly influential in encouraging families to acquire broadband service. According to his research among at-risk group population in the United States, 98 % of families said they got broadband service because their children needed it for school (91 % said their children influenced their decision, will 60 % said they were influenced by their children's teachers).

As the examples below will show, successful digital literacy programs in primary education tend to focus holistically on the provision of computers to students, the subsidization of broadband service both for students via Wi-Fi and for schools via fixed facilities, and the intense embedding of ICT programs in the formal curriculum. In many cases, successful initiatives in this domain are associated with partnerships between the public and private sectors.

Case Study 3.1: "Plan Ceibal" (Uruguay)

Plan Ceibal initially aimed to provide every primary school student with a computer as part of the One Laptop Per Child (OLPC) initiative and now includes secondary school students. Plan Ceibal also brings Internet to the schools and incorporates IT training into the curriculum. It launched in 2006 to address the country's digital divide. By providing students with laptops that they can take home, ICT access increases not only amongst the students, but also within their families as well. Indirectly, the provision of laptops also addresses national digital literacy; as students become familiar with the technology in the classroom, they then share this knowledge with family members, which then spreads to other members of the community.

By 2011, total investment in the program amounted to approximately US$ 100 MN, or 0.25 % of the country's GDP and 8 % of its education-related expenditures. This figure covers the US$ 250 spent per student, derived from the US$ 188 cost of the laptop and the US$ 60 in maintenance and Wi-Fi connection fees over 4 years. By mid-2012, nearly 600,000 Uruguayan students owned personal computers and 99 % of the nation's 4,375 primary and secondary schools had Internet access.

An August 2013 study, however, shows that despite this increased computer access, the initiative has not impacted students' math or reading scores. The next step, then, would be to ensure (a) that teachers understand technology's potential to improve the learning process, and (b) that they have the proper training to integrate it into their lesson plans.

Sources

Prusa, Anna, and Elizabeth Plotts. "Uruguay's Plan Ceibal: Can Laptops in the Hands of Primary School Students Reduce the Digital Divide, Improve Education, and Increase Competitiveness?" *Capstone Project*. George Washington University, 26 Apr. 2011. http://elliott.gwu.edu/assets/docs/acad/lahs/uruguay-potts-prusa-2011.pdf.

"Uruguay: Plan Ceibal Ends Digital Gap in Public Schools." *InfoSur Hoy*. N.p., 10 Apr. 2010. Web.http://infosurhoy.com/cocoon/saii/xhtml/en_GB/features/saii/features/society/2012/04/10/feature-02.

"One Child-one Laptop" Program in Uruguay Has Failed to Improve Results in Maths and Reading." *MercoPress*. N.p., 19 Sept. 2013. Web. http://en.mercopress.com/2013/09/19/one-child-one-laptop-program-in-uruguay-has-failed-to-improve-results-in-maths-and-reading.

Case Study 3.2: Education Modernization Program (Russia)

The Russian government has included ICT training in its education system since 1986, though until recently this training focused on secondary and higher education. In 2006, the government connected all public schools to the Internet, and with this push to increase access to technology came the introduction of ICT training to primary school classrooms. The Federal Education Agency of Russia recommended that schools develop a formal computer curriculum, encouraging the use of ICT as part of the curriculum from Grade 2—or age seven—onward.

Beyond simply increasing the presence of ICT in the classroom, the Russian government partnered with the World Bank's Russia office in 2004 to implement the "Education Modernization Program," which supported the improvement of ICT skills and competencies through e-learning. While the program covered all levels of education and government services, it specified that 20 % of resources must be used toward ICT at the primary school level.

The program was developed to address the disparities in ICT access and competency throughout the country. As Russia transitioned to a market economy, many regions inherited failing education systems and did not receive adequate funding to provide students with marketable skills for employment. Per the 2000 OECD PISA assessment, Russian students ranked 27th of 31 countries. The government concluded that ICT competency would improve the quality of the workforce, and that ICT could enhance the access and quality of its education system.

Once the government provided schools with basic computers and Internet connections the three phases of the 4-year e-learning project included: "(i) development of new learning materials; (ii) support for both pre-service and in-service teacher training in the introduction of ICT into teaching and learning; and (iii) establishing not less than 200 resource centers to improve access to ICT enhanced education opportunities and to disseminate new teaching practices." Additionally, by the end of 2007, all schools had broadband connections.

The entire cost of the project totaled US$ 145 MN, broken into US$ 35 MN on learning materials, US$ 43 MN on teacher training, US$ 63 MN on resource centers, and US$ 4 MN on project management.

To measure efficacy, the program incorporated a series of 18 indicators, including: teacher competency in Internet education, the incorporation of digital resources into the classroom, and the creation of open access textbooks. The program met or exceeded each of these goals. By completion, the number of students enrolled in distance learning had increased by 75 %, with the number of rural students accessing online education multiplied by 5. Further, ICT competency levels improved and the availability of e-resources grew.

Sources

"Education Transformation in Russia." *Intel*. N.p., 2009. Web. http://www.intel. com/content/dam/doc/case-study/learning-series-education-transformation-study.pdf.

"E-Learning Support Project." *Education*. The World Bank, 30 Dec. 2008. Web. http://web.worldbank.org/external/projects/main?pagePK=64283627.

"Implementation and Completion Results Report." The World Bank, 30 Dec. 2008. Web. http://www-wds.worldbank.org/external/default/WDSContent Server/WDSP/IB/2009/01/13/000333038_20090113000324/Rendered/PDF/ICR10100ICR0Bo1dislcosed0Jan0902009.pdf.

Case Study 3.3: National ICT Literacy Assessment (Australia)

In 2008, Australia introduced measures to assess students' digital literacy skills and comprehension. Students in years 6 and 10 (ages 12 and 16 on average) sat for standardized tests, which ranked digital literacy comprehension on a 6-point scale. The lowest level, Level 1, designated a basic understanding of how to use a computer and software. The highest level, Level 6, was reserved for students who could use advanced software features to organize information, synthesize data, and complete information products. Prior to this assessment, the country had testing in place to identify areas of weakness in its traditional literacy and numeracy education, but not in its ICT curriculum.

To develop the test, Australia's Ministerial Council for Education, Employment, Training and Youth Affairs (MCEETYA) partnered in 2005 with the Australian Council for Educational Research (ACER). The independent non-profit organization produced the National Sample Assessment of ICT literacy, the first of its kind. The initial trial tested 620 Australian students and evaluated students' analytical skills rather than simply their software know-how.

Per the results of the 2008 testing, more than 40 % of year 6 students tested at a Level 3 or higher, meaning that they could conduct simple Internet searches and identify the best source. At almost half of all year 10 students tested at a Level 4 or higher, meaning that they could conduct more complex searches and use the information they found to generate new content.

In 2011, the OECD released its results from the Electronic Reading Assessment (ERA) component of the 2009 OECD Program for International Student Assessment (PISA). The overall PISA examination tests high school students' readiness to enter and contribute positively to society following the end of their compulsory education. The exam places high value on the ability to address "real world" situations rather than on specific curriculum items. Within this context, the ERA specifically looks at students' ability to navigate through electronic text, as developed through exposure to ICT. Of the 19 countries surveyed, Australia ranked second in terms of the ERA.

The PISA 2015 assessment will not include the digital reading component, but a computer-based assessment will be the primary mode of testing. Countries will have the option, however, to choose not to test students by computer.

Sources

"Broadband Strategies Handbook." Ed. Tim Kelly and Carlo M. Rossotto. The World Bank, 2012. Web. https://openknowledge.worldbank.org/handle/10986/6009.

Ainley, John. "Measuring Australian Students' ICT Literacy." *Research Developments* 14.5 (2005). *ACER*. Web. http://research.acer.edu.au/resdev/vol14/iss14/5/.

"OECD Programme for International Student Assessment (PISA)." ACER, 2012. Web. http://www.acer.edu.au/ozpisa/assessment/.

3.1.1.2 Secondary School Programs

While secondary schools do address basic digital literacy skills, they tend to provide students with a more advanced knowledge than they would gain during their primary school years. Viewing digital literacy as a life skill can explain its application in a student's life well beyond the classroom. As the global economy shifts from the manufacturing of goods to the provision of services, workers and countries require more high-level skills to stay competitive. Particularly in instances where students move directly from secondary or vocational school to the workforce, the exposure they have to ICT training via the education system has the potential to shape the trajectory of their future careers and the strength of the national economy. Employers increasingly require digital competence, and workers with this type of training also tend to acquire other on-the-job skills more easily. Further, the ICT industry tends to offer more high paying, lucrative jobs, adding financial incentive to the benefits of obtaining advanced digital literacy.

By incorporating digital literacy training into the secondary school system, policy makers can effectively bridge the digital divide, thus creating more equal workforce opportunity amongst the population. Further, employees comfortable with using the technology at work are more likely to see its value within the household.

Given that most countries now require secondary school attendance, this environment seems to serve as the ideal setting in which to introduce citizens to basic and advanced ICT training. Training cannot come to fruition, however, without the necessary technology. In addition to developing effective and applicable lesson plans, educators and policy makers must also consider the provision of personal computers coupled with broadband connectivity. To this end, an increasing number of government initiatives have focused on distributing laptops to secondary students and faculty members. Some governments, such as North Carolina in the United States, require students to pass an ICT competence exam in the seventh or eighth grade to receive a high school diploma.

As is the case with primary school digital literacy programs, educators should have some form of measurement or standardization in place to promote the efficacy of such initiatives. Successful examples have included testing, certification programs, and partnerships with international organizations. As in the case of primary school programs, successful initiatives are also based on public and private partnerships.

Case Study 3.4: Conectar Igualdad and Educ.ar (Argentina)

In an effort to promote digital literacy in the country, in 2010 Argentina established its national Conectar Igualdad program. The first phase of the program targeted the country's secondary schools, promising equal ICT access to all students in urban and rural areas.

While Conectar Igualdad aimed to distribute 3 million laptops to secondary students and teachers, it recognized that access alone would not increase digital literacy. Beyond laptop distribution, the initiative also included Internet access and internal networks within the schools, the creation of digital content, and a standardized program to train teachers on how to incorporate ICT use into the classroom. To this end, Conectar Igualdad complemented the country's *Educ.ar* platform, which was designed to assist teachers in the development of an ICT curriculum by creating a standard set of materials for use throughout all schools.

In developing Conectar Igualdad, the Argentine Ministry of Education partnered with other sectors of the government—the Social Security National Administration, the Ministry for Federal Planning, Public Investment and Services, and the Executive Branch Cabinet's Head. By incorporating these high-level agencies, the program promoted centralization and discouraged an uneven distribution of resources.

By purchasing computers on such a large scale, each unit costs approximately US\$ 350 (as opposed to the average US\$ 506). While the government covers the connectivity and training costs, *Educ.ar* operates as a private enterprise and relies on pro-bono support from educational institutes and corporations.

By May 2012, Conectar Igualdad had so far distributed 1.8 million laptops and *Educ.ar* had created more than 20,000 pieces of material specific to the secondary school curriculum. By the end of 2013, the program had reached 3.6 million students and faculty members from secondary schools and specialty schools, with universal access slated for 2014. Together, through increased ICT access and formal training, Conectar Igualdad and *Educ.ar* effectively promoted digital literacy throughout Argentina's secondary schools.

Sources

Finquelievich, Susana, Patricio Feldman, and Celina Fischnaller. "Public Policies on Media and Information Literacy and Education in Latin America: Overview and Proposals." Proc. of Media and Information Literacy in Knowledge Societies, Atlas Park Hotel, Moscow. N.p., 28 June 2012. Web. http://www.academia.edu/1795719/_Public_Policies_on_Media_and_information_literacy_and_education_in_Latin_America_Overview_and_Proposals_.

"Conectar Igualdad." *La Presidenta Anuncio La Adjudicacion De 1,5 Millones De Netbooks*. N.p., 18 Feb. 2011. Web. http://www.conectarigualdad.gob.ar/noticias/actos/la-presidenta-anuncio-la-adjudicacion-de-15-millones-de-netbooks/.

Giangola, Norissa. "What Works: Educ.ar's Strategy for a Nation Connected and Learning." World Resources Institute, July 2001. Web. http://pdf.wri.org/dd_educar.pdf.

"Conectar Igualdad "completará El Universo De Estudiantes" En 2014—Télam—Agencia Nacional De Noticias." *Conectar Igualdad "completará*

El Universo De Estudiantes" En 2014—Télam—Agencia Nacional De Noticias. N.p., 21 Nov. 2013. Web. http://www.telam.com.ar/notas/201311/41792-conectar-igualdad-completara-el-universo-de-estudiantes-en-2014.html.

Case Study 3.5: ICDL Accreditation (Senegal)

Designed to serve as an international standard for computer competency, the International Computer Driving License (ICDL) certification program was initially developed for use in European nations by the Council of European Professional Informatics Societies (CEPIS). Its success spread, however, and citizens in 48 countries now have access to the program. In order to receive the license, candidates must pass a series of tests on modules covering various ICT-related subjects ranging from word processing to web browsing. In preparation, individuals typically take a training course before sitting for the 45-min exam.

In 2010, Senegal's Ministry of Vocational and Technical Training established a formal partnership with ICDL—Africa and USAID to ensure a standardization of digital literacy training within the country's schools. The project introduced the certification program to students in the country's middle schools with the goal of certifying all students within four years.

When announcing the program, the minister of Vocational and Technical Training emphasized the potential for ICT skills to impact national development and close the socio-economic divide by potentially empowering otherwise disadvantaged groups. By partnering with ICDL, the Ministry can ensure effective quality digital literacy training. All holders of the certification have demonstrated an ICT and digital literacy competency. With the support of national governments, private corporations, and international organizations, it is now internationally recognized and available in 148 countries. To date, there are more than 11 million global candidates, making the ICDL the world's largest end-user computer skills certification program.

The first step in implementation in Senegal involved a pilot testing of 100 students, which led to the accreditation of select institutions to serve as official training and exam centers. Following this initial phase, universities and other educational facilities joined and the program eventually expanded to include partnerships with corporations to assist with publicity and funding. Within a year, 113 middle schools registered to participate. As written into the program, local citizens run all management and training.

In the event that students cannot attend class, they have the option to download the ICDL syllabus free of charge. The examination requires a "small cost," and supplemental training courses cost extra. While the program itself does not yet offer scholarships, corporate sponsors may do so in the future.

As of 2014, more than 12 million people across the world had accessed ICDL certification programs in 41 languages at more than 24,000 test centers.

Sources

"ECDL Foundation." *European Computer Driving Licence Foundation.* N.p., n.d. Web. http://www.ecdl.org/index.jsp?p=94.

"ICDL Africa." N.p., n.d. Web. http://www.icdlafrica.org/index.jsp.

"Licence to Skill." *ELearning Africa News Portal.* N.p., 15 Jan. 2009. Web. http://www.elearning-africa.com/eLA_Newsportal/licence-to-skill/.

Case Study 3.6: GILAS (Philippines)

Gearing up Internet Literacy and Access for Students, or GILAS for short, provides public secondary schools in the Philippines with computer labs, complete with Internet connections, software, basic hardware, and Internet training. The initiative, which began in 2005, is a partnership between 26 corporations and non-profit institutions that recognize the limitations of the government's education budget. Amongst other goals, the project aims to deliver Internet access and computer equipment for schools, training for teachers and administrators, and the formulation of lesson plans.

By providing Internet access to schools, sponsors of the project see it as a means of bridging the digital divide among public high school students. Only a small number of Filipino students attend college, largely due to the prohibitive costs of higher education in the country. Beyond a college education, many employers see Internet literacy as a hiring requisite, though this skill is typically reserved for wealthier students whose families can afford household computers and Internet connections. Without computer access or the ability to afford a college education, many students have few opportunities awaiting them at graduation. By increasing computer access and digital literacy within the school system, the GILAS project aims to produce a more qualified and highly skilled workforce.

To support the initiative, GILAS matched donations from local and foreign companies, local governments, and legislators. Per the most recently released annual report, the public sector 2009 contribution added up to approximately US$ 500,860 in addition to the private sector's US$ 598,470 contribution. Overseas Filipino expatriates also made donations, mainly through the Ayala Foundation USA, that totaled US$ 175,980. In total, donations that year equaled US$ 1.3 million.

In 2010, the country's Department of Education initiated its DepEd Internet Connectivity Project (DICP) with the intention of connecting all public high schools to the Internet while providing relevant monitoring through an annual allocation of US$ 1,200 per school. The initiative complemented the GILAS program and leaders of both projects worked together to reach their shared goal. DICP focused more on financing schools' Internet connections while GILAS looked more at the initial investment in the provision of ICT tools and training.

Within 4 years of its 2005 inception, the GILAS program connected 39 % of the Philippines' public high schools. As a result, more than 2 million students accessed the Internet and 11,621 teachers received training. By late 2012, the program had reached a total of 3,349 schools.

Sources

GILAS: Gearing up Internet Literacy and Access for Students, n.d. Web. http://www.gilas.org/.

2009 Annual Report: On the Way to Sustainability. Rep. GILAS: Gearing up Internet Literacy and Access for Students, 2010. Web. http://www.gilas. org/attachments/AR_2009.pdf.

"DICP." *DepEd Division of Malaybalay City*. N.p., n.d. Web. http://www. depedmalaybalay.net/programs/ict/deped-internet-connectivity-project-dicp.

3.1.1.3 Distance Learning Programs

The introduction of technology in education changes the scale in the delivery of the educational product, both in terms of resource availability as well as regarding the training of teachers and professors. In addition, by its nature, technology has the potential to break the distance barrier, becoming a fundamental tool to meet the needs of population in remote areas. The use of ICT in education can generate a significant contribution in four areas:

- Learning support to students in quantitative subjects such as geometry, basic algebra and hard sciences;
- Teaching support in regions where achievement tests yield results that are lower than the national average, whereby they might be linked to socio-economic, ethnic, or gender gap;
- Contribution to address shortfalls in adolescent students that exhibit reading and comprehension difficulties;
- Resource support for teachers;

Finally, ICT also appears to have a potential, indirectly through some of the effects mentioned above, or directly, to reduce student attrition, particularly in areas where rates approach 30 %. Having said that, the return on the technology investment in education in remote areas depends on three factors: the contents delivered through technology have to be adapted to the technology format, the processes and principles guiding the teaching experience have to be redesigned in order to incorporate the technology input, and finally, teachers and professors have to be trained on the utilization of technology. These three requirements are of paramount importance in order to ensure that the technology investment in remote areas yields the expected results. In the first condition, research indicates that educational contents need to be adapted to the new teaching formats, rather than being merely copied and digitized. Under the second requirement, the teaching process has to be transformed in order to render the use of technology within a new context of enhanced learning that blends the classroom experience with the technology support. Finally, teachers and professors need to be trained so they can not only familiarize themselves with the technology but also learn how to use it effectively in the classroom.

In light of these conditions, technology plays a very important role in the transformation of the educational institution located in remote areas. In the first place, the technology platform becomes a learning tool inserted within a learning experience that is student- focused. Secondly, technology becomes a vehicle for delivering resources and teacher support. Third, technology becomes an enabler to facilitate the transition of students in an information society.

There are several broadband-enabled technology platforms that can contribute in terms of the benefits pointed above:

- Video programs distributed through broadband: there is considerable research supporting the educational value of distance learning through television as a complementary resource to the classroom. With the introduction of broadband and the use of computers, the development of "learning objects", which constitute small video segments that can be easily integrated with mathematics, history and geography curricula, have become commonplace. These programs have also started to be modified in order to be able to be distributed through smartphone screen formats;
- Interactive whiteboards: despite being a new technology, these tools can play a very positive role with regards to the use of methods promoting student engagement, that can be easily adapted to different learning styles;
- Portable terminals, such as personal computers, tablets and smartphones can address limitations in terms of access to content, promoting student independence in those educational settings that have a small number of teachers (such as schools with a single teacher in remote areas). However, this effect can be fulfilled if content is adapted to the different screen formats, which in some cases, can be fairly small;
- Virtual learning platforms: these technologies comprise traditional audiovisual tools, such as videoconferencing, or more o advanced based on the Internet, with a capability of operating in real time with an upstream channel. These types of tools can very useful in teaching semantics and conceptual mapping (such as story lines, and roadmaps to structure the writing of essays), geometry and hard sciences.

The contribution of technology in education remote areas covers numerous applications domains from distance learning to the utilization of portable terminals and computer-based tools. In the case of distance learning, research shows that students in remote areas whose primary vehicle of instruction is the technology platform tend to perform at an equivalent level that those students learning in traditional classrooms (Hudson 2006). The studies of Witherspoon et al. (1993) also show that, despite the distance and physical separation from the teacher, the students undergoing a distance learning program tend to be more motivated to learn, and are more mature. On the other hand, the design of educational material tends to be more systematic and oriented toward making the learning process more efficient. Finally, Hudson indicates that distance-learning programs combined with tutoring delivered via satellite tend to reduce school absenteeism (Hudson 1990).

Another important effect of technology in education is the impact of the Internet. While the extent of its contribution is highly dependent on the quality of equipment and the telecommunications access, when those factors are controlled in study settings, the Internet can compensate for variables such as low training of teachers or lack of educational material (Puma et al. 2000). Social networks, which citizens access over broadband, can deliver education to students in remote areas.

Case Study 3.7: MOOCs: Coursera and EdX (United States and International)

In 2008, researchers and professors from the National Research Council of Canada and the University of Manitoba launched a course on learning theory, and opened it to 25 tuition-paying university students as well as to 2300 members of the public who could take the course—for free—online. Dubbed a "massive open online course," or MOOC, this concept now includes countless free web-based classes that are designed to reach a large audience and stress convenience and individual learning pace. At present, two large American ventures dominate the MOOC environment—Coursera and EdX.

Founded by two Stanford University computer professors in 2012, by 2014 the education company Coursera has grown into a partnership with 108 global universities, offering 618 free online courses to more than 5 million students. As described by Coursera co-founder Andrew Ng, this project can "bring higher education to places where access is limited, and… give established educational institutions opportunities to raise their impact both on and off campus."

Founded and governed by the Massachusetts Institute of Technology (MIT) and Harvard University (both also in the United States) in May 2012, the non-profit EdX has expanded to include 30 global university partners, although more than 200 institutions expressed interest in collaborating in its 1st year. The program started when one professor offered his electrical engineering course online; by early 2014, it offered 146 courses. EdX aims to eliminate traditional barriers to quality education access for students all over the world, including age, income, nationality, and location. By 2025, EdX expects to have worked with one billion students across the world.

From their inception, both Coursera and EdX have seen a large international student presence, and their emerging partnerships with international universities will likely only increase this trend and demand for online university services.

While the programs will continue to offer their online courses for free, they are both pushing a new model that will offer students formal college accreditation, but at a fee. To do so, they will soon incorporate such tools as identity-verified certificates, proctored exams, and recommendations from the American Council on Education, all of which many universities look for when considering transfer credit.

Sources

Tamburri, Rosanna. "All about MOOCs." *University Affairs*. N.p., 7 Nov. 2012. Web. http://www.universityaffairs.ca/all-about-moocs.aspx.

"Mooc Definition." *Financial Times Lexicon*. N.p., n.d. Web. http://lexicon.ft.com/Term?term=mooc.

Lewin, Tamar. "Universities Abroad Join Partnerships on the Web." *New York Times*. N.p., 20 Feb. 2013. Web. http://www.nytimes.com/2013/02/21/education/universities-abroad-join-mooc-course-projects.html?_r=0.

"About Coursera." *Coursera*. N.p., n.d. Web. https://www.coursera.org/about.

"About EdX." *EdX*. N.p., n.d. Web. https://www.edx.org/about.

Lorch, Kristina D., and Conor J. Reilley. "The Harvard Crimson." *EdX: Expanding Education Beyond the American University | News |*. N.p., 18 Dec. 2013. Web. http://www.thecrimson.com/article/2013/12/18/edX-global-expansion-degrees/?page=1.

The benefits of technology in education can materialize when coupled with teacher training to break down adoption structural constraints. Similarly, technology training must extend to the mid-level functionaries of ministries of education, where the impetus for the introduction of technology tools often resides. Ultimately, one of the most important challenges in this domain is transforming the culture and resistance to technology in educational institutions and ministries of education.

Case Study 3.8: Center for E-Learning and Distance Education (Saudi Arabia)

Saudi Arabia created a taskforce dedicated to the preparation of an e-learning and distance education model to create more equal educational opportunity. Distance education addresses physical location and time constraints, creating opportunities regardless of students' age, gender, or lifestyle. As education quality and availability improves, so too should the country's labor market. In 2013, the government allocated US\$ 54.4 billion—25 % of its federal budget—to education and training. By 2026, it plans to spend 2 % of its GDP on research, a figure comparable to the United States and the United Kingdom.

In 2006, the Ministry of Higher Education contracted the Malaysian consortium METEOR to create the Center for E-Learning and Distance Education (NCeL) for US\$ 12.5 MN. The Center now serves as the hub for university e-learning and distance education programs, providing resources and training to create a more accessible education program while addressing the country's shortage of qualified professors. The center also sets the standards for the design and production of materials while coordinating with international higher education institutions. By year-end 2013, NCeL offered more than 100 digital courses and 200 research studies to 400,000 students, faculty members, and researchers.

The Center's initiatives have included the creation of an educational portal that facilitates the exchange of e-learning education-related experiences. The portal includes a forum where users can discuss their experiences and share their expertise in the field as well as a news section for a compilation of all information pertaining to NCeL and the field of distance education. Because the materials are all developed as open source, students can easily access and share them. Similarly, the Center also established the Saudi Digital Library, which now offers 260,000 digital books to all university students and faculty.

Sources

E-Learning and Distance Education. Publication. Ministry of Higher Education, 2011. Web. http://www.icde.org/filestore/Resources/Handbooks/e-LearningAndDistanceEducation.pdf.

"Saudi Digital Library and Nature Publishing Group Sign Agreement." Nature Publishing Group, 27 Nov. 2013. Web. http://www.nature.com/press_releases/saudi-digital.html.

"Active Local Roles…with International Presence." The National Center for E-Learning and Distance Learning, 24 Dec. 2013. Web. http://www.online-educa.com/cms7/sites/default/files/documents/downloads/national-center-for-e-learning-and-distance-learning.pdf.

3.1.1.4 Digital Literacy for Teachers

Digital literacy programs focusing on teachers represent a critical component of formal education changes. Any change in the formal curriculum in primary and/or secondary education that is not complemented with teacher retraining has a high failure rate. At best, given the digital awareness of young cohorts, these situations can result in the syndrome known as "digital children/analogue teachers," whereby teachers are not capable of providing the necessary training and/or interacting with children on the basis of ICT usage.

Many of these initiatives go beyond increasing teachers' digital literacy, offering resources and instruction on the incorporation of ICT training into the classroom. Naturally, the heightened exposure to these tools makes teachers more comfortable with the technology and more aware of its benefits in and out of the formal education system. Aside from schooling students to use computers and the Internet effectively, many teachers also introduce online learning supplements—such as games, assessments, and e-books—that require interaction and fast speeds to enhance the education of their students. As students become more engaged and technology use becomes second nature, they bring these skills home with them, unknowingly passing their knowledge along to other members of their families and communities.

As is the case with primary and secondary school digital literacy initiatives, the training of teachers requires an investment not only in the instruction but also in the relevant technology. Without access to computers, related software, and broadband, these programs would have little to no value or application. Various initiatives address this issue differently, though many successful programs tend to provide both students and teachers with laptop computers. Local and national government partnerships with international corporations, NGOs, and multilaterals offer both funding and quality instruction and resources that incorporate the best experiences of past projects.

The implementation of teacher training varies, but the cost efficient "train the trainer" model appears to have the best results. In this instance, project leaders work with select teachers—through online training courses, intensive workshops, or certification programs—who then train other teachers and faculty members. Some programs start with urban schools and then expand to rural schools, while others target schools with the most need. Others still selectively choose schools based on their geographic proximity to other schools and resources.

Ultimately, an investment in teachers is an investment in human capital. High-skilled teachers produce high-skilled students, which lead to large-scale and long-term sustainable economic growth.

Case Study 3.9: Microsoft IT Academy (Nigeria)

The Microsoft IT Academy Program—which now issues over 1 million certifications per year to students in more than 160 countries worldwide—is designed to provide students with the ICT training necessary to stay competitive in the workforce. Beyond courses and certifications for students, it also offers resources for teachers and faculty members that include training, lesson plans, E-learning, student projects, and assessments. By providing such tools, teachers can more easily incorporate new technology into their classroom curricula and effectively use them to create specialized age appropriate lesson plans. The Academy allows educators to access to its e-reference libraries as well as its database of resources designed specifically for educators and students looking for advanced IT-training.

In the summer of 2010, Microsoft Nigeria partnered with the Lagos State Government to bring the IT Academy to local secondary schools in Agidingbi as part of its efforts to modernize public education. As part of this pilot, Deux Project Limited—a Nigerian company providing construction and consulting services—worked with Microsoft to provide Microsoft software-equipped laptops to the schools. The partnership was recognized as means to address the country's lack of quality training and a resource for teachers looking to supplement their course plans with online learning tools.

Deux Project Limited managed the academy while working closely with Microsoft to ensure the fulfillment of its requirements. The Academy offered students more than 175 e-learning courses and email addresses while increasing faculty access to online resources.

In January 2012, the Microsoft IT Academy concluded the pilot phase of its teacher-training program, Digital Literacy Curriculum for School Teachers. During this 2-week training, teachers learned how to utilize ICT to teach their students to do the same. This initial pilot drew 83 teachers, which in turn led to the classification of their schools as "Certified Microsoft IT Academies."

Sources

"IT Academy Program Overview." *Microsoft IT Academy Program*. Microsoft, 2012. Web. http://www.microsoft.com/en-us/itacademy/overview.aspx.

"Microsoft, Lagos Partner on IT Academy Programme." *Nigeria News*. N.p., 15 June 2010. Web. http://news2.onlinenigeria.com/index.php?news=33176.

"Pilot Training in Digital Literacy Curriculum Concluded in Lagos State." *Lagos Indicator*. N.p., 29 Jan. 2012. Web. http://www.lagosindicatoronline.com/Pilot_Training.html.

Case Study 3.10: Intel Teach Program (Sri Lanka)

Since 1999, the international Intel Teach Program has trained over 10 million teachers in 70 countries. The training focuses on providing teachers with the professional development needed to integrate technology into their classrooms and improve their curricula. Intel developed its education model through partnerships with governments, NGOs, multilateral organizations, and educators and research spanning more than a decade. Ultimately, it created a program emphasizing five main facets: policy reform, curriculum and assessment, teacher profession development, ICT, and research and development. Based on these five components, Intel structures its programs to address each country's specific education needs.

In 2006, Intel signed a Memorandum of Understanding with Sri Lanka's Ministry of Education to implement a teacher development program designed specifically for the needs of the Sri Lanka education system at no cost to the government. The project introduced a variety of tools including "skoool Sri Lanka," an interactive web-based program promoting math and science learning. Skoool offers students practice exams and instruction tailored to their individual areas of difficulty. In order to use the program, classrooms must have access to a PC with Internet capabilities. The Intel Teach Elements website offers Sri Lankan teachers lesson modules that promote Project-Based Learning. When teachers log in, they receive help in organizing their curricula to encourage training relevant to the demands of the twenty-first century.

One year after its inception, Sri Lanka organized the "South Asia Intel Teach Program Forum" for policy makers from the country as well as from India and Pakistan to promote classroom ICT integration in the region. In November 2008, the Ministry of Education hosted an awards ceremony to recognize the first 100 "Master Teachers" who had completed the Intel Teach training program. By 2013, 40,000 teachers in Sri Lanka had received training through the program.

Sources

"Intel Teach Program Worldwide." *Intel*. N.p., n.d. Web. http://www.intel.com/content/www/us/en/education/k12/intel-teach-ww.html.

"About Skoool." *Skoool Sri Lanka*. N.p., n.d. Web. http://www.skoool.lk/about_skoool.htm.

"Intel Teach Elements: Project-Based Approaches." *Intel Education Initiative, Sri Lanka*. Intel, n.d. Web. http://www.intel.com/cd/corporate/education/apac/eng/lk/tools/elements/446491.htm.

"Inside WSA Global Congress 2013—Day 03." *Read Me*. N.p., 25 Oct. 2013. Web. http://readme.lk/wsa-global-congress-2013-day-03/.

Case Study 3.11: Regional ICT Training and Research Center (Rwanda)

In 2004, Rwanda's Regional ICT Training and Research Center (RICT) was established to strengthen the population's ICT skills to support sustainable economic growth. The center provides students with basic training as well as the skillset needed to enter the IT industry. This emphasis on computer literacy fell in line with the Ministry of Finance and Economic Planning's "Vision 2020," which included initiatives to prepare the country for a transition to a knowledge-based economy by 2020. Amongst other facets, Vision 2020 addressed the need to support ICT skills aimed at public sector, private sector, and civil society as well as the development of ICT network infrastructure.

In 2009, RITC announced plans to ensure computer literacy amongst all Rwandan teachers by 2010, with an emphasis on hands-on computer skills building. Phase one targeted 5,000 primary school teachers and the second phase reached secondary school teachers. While the project initially focused on teachers in urban schools, it later expanded to reach the needs of the rural teachers. The plan also included the distribution of 100,000 XO computers to schools.

Prior to this push, Microsoft Partners in Learning (PIL) worked with RITC to provide 3,000 secondary school teachers with basic ICT skills. The project took place in 2005 and received funding from the Ministry of Education. The following year, RICT conducted an in-depth ICT training project for 1,000 secondary school teachers using the same model. Both projects followed a "trainer-of-trainers" model, whereby two teachers from each school received higher-level instruction to then train other teachers and act as troubleshooters. This project had a budget of US\$ 129,540. Microsoft PIL pledged an initial US\$ 37,900 and the Microsoft Emerging Markets Team contributed an additional US\$ 25,000. RITC contributed US\$ 15,750, with the Ministry of Education covering the remaining US\$ 50,890.

Sources

Regional ICT Training and Research Centre, 2008. *RITC Prospectus 2008.* Kigali: RITC

"Rwanda Vision 2020." MINECOFIN, n.d. Web. http://www.minecofin. gov.rw/ministry/key/vision2020.

Farrell, Glen. *ICT in Education in Rwanda.* Rep. InfoDev, Apr. 2007. Web. www.infodev.org.

ICT in Education Support Initiatives. Rep. Ministry of Education, Science, Technology, and Scientific Research, n.d. Web. http://www.docstoc.com/docs/48796695/ICT_IN_EDUCATION_SUPPORT_INITIATIVES.

3.1.2 Targeted Digital Literacy Programs

While digital literacy embedded in formal education processes are conducted in school institutions, closely linked to curricula, targeted programs entail group-specific training in the use of computers and broadband typically delivered through a range of public access centers. This section reviews the major categories of targeted programs.

Targeted digital literacy programs are of a wide variety, potentially addressing a number of objectives, not all necessarily consistent. In designing such programs, policy makers need to consider what are the goals of the program, since these goals will frame the methods of intervention. Among the goals to be considered in designing a digital literacy program, the following issues need to be addressed:

- What is the overall objective of the program? Digital literacy, conceived as a skill, represents the means to achieve a varying set of goals, such as improvement of quality of life, develop citizenship and promote democratic participation, or social inclusion. By outlining the ultimate objective, policy makers will help framing the program.
- As expected, digital literacy programs could have more than one objective, partly driven by the population being targeted. For example, if targeting the rural poor, the purpose of the digital literacy program could include providing access to broadband, improving quality of life to prevent rural exodus to cities, and promoting social inclusion. As Hilding-Hamann et al. (2009) mention in their report to the European Commission, the differences in program objectives could "reflect different policy domains" (e.g. education, economic development, social welfare). The potentially different constituencies sponsoring the program could also drive program objectives.
- Formal versus informal delivery mode? Formal digital literacy training entails structured programs based on established curricula, learning tools, and certification. Informal training is not delivered in specific training environments, lacking a structured pedagogical process. While it might not be intuitively appropriate for targeted programs, the emergence of new Internet platforms might lead to the adoption of informal approaches.
- Scale of implementation? This question addresses whether programs will be focused on a particular region, or deployed on a national scale. In Hilding-Hamann et al. (2009) view, "national programs are rooted in centralized policies at the national level and (...) seen as strategically linked to government objectives", such as building an information society. In general terms, local programs, while having a more limited impact across targeted populations, tend to experience a large sustainability success rate due to more limited funding requirements. Nevertheless, Hilding-Hamann et al. (2009) did not find a relation between size of the program and sustainability.
  Sustainability is a primary concern of targeted digital literacy programs. In their review of 464 programs, Hilding-Hamann et al. (2009) estimated that 22 % of them had been discontinued. Furthermore, they found that program sustainability is generally linked to the number of stakeholders ("more than half of the (ongoing) initiatives have been delivered by three or more implementers").
- What is the target group? Targeted digital literacy programs take different shapes according to the population they will address. As an example, the type of content to be emphasized in program delivery will change significantly if the program aims to target the elderly (teaching email for social inclusion and

fostering of social and family ties) versus adults (training on applications to build employability skills). It is often the case that even needs within a single targeted group might be of different types. For example, some digital literacy programs that target the elderly have focused on helping users work with specific devices, while others have focused on basic operations and routines of operating systems.

- Usability versus accessibility? Some digital literacy programs emphasize training and skills transmission, while others complement this with infrastructure for public broadband access. This represents a critical policy choice since access does not necessarily equate to the capability to use broadband in a productive and beneficial manner. In fact, if the primary objective is usability, experience indicates that tailored courses, complemented with intense coaching, are the more appropriate approach.

  As expected, if the target of the digital literacy program is the rural poor, accessibility will be a dominant objective. A combination of both objectives – use and access- can be provided by community access centers, which will be reviewed later. Nevertheless, best practices indicate that accessibility and usability are not that easy to combine in digital literacy programs. As such, the two objectives are frequently addressed sequentially, first providing access, followed by training.

- Device focus: Until now, the great majority of digital literacy programs have focused on personal computers connected to broadband technology. However, with the growing importance of wireless broadband and smartphones, the need to make decisions on what kind of device the digital literacy program focuses on will become very important.

In the 10 year period between 1998 and 2008, the number of domestic IT jobs in the United States increased 26 %, compared to just 6 % of overall employment growth. Nearly all Americans (96 %) now use ICT daily, and the majority (62 %) uses the Internet as part of their jobs. Further, these skills allow citizens to search and apply for these jobs and promote access to other valuable resources such as online college courses and government services. Digital literacy ultimately improves not only the employment opportunities for individual citizens, but the country's competitiveness and economy as well. In the United States, for instance, Internet-related jobs in the country added US$ 300 billion in economic activity to its GDP in 2009.[1]

Other developed countries experience the same need for employees with digital know how. In the UK, for instance, 90 % of jobs require "some level of IT competency." With more than 10 million Internet users in the country, those citizens without access or digital literacy skills will soon find themselves "even more isolated and disadvantaged," particularly as every day services move online.

---

[1] "Fact Sheet: Digital Literacy." *United States Department of Commerce*. N.p., 13 May 2011. Web. 08 Mar. 2013. http://www.commerce.gov/news/fact-sheets/2011/05/13/fact-sheet-digital-literacy.

As such, the larger education system and national initiatives should focus on the provision of digital literacy training at all levels. Students must leave the class-room ready to enter a world requiring IT skills, while adults must have access to necessary instruction to keep pace with the skill-biased technological change. Incorporating such programs into the education system ensures sustainability and funding. As a basic skill—much like traditional literacy and numeracy—digital literacy ought to be included in all areas of the formal curriculum. Further, numerous studies have shown a link between digital literacy and excellence in other academic areas, concluding that technology use in the classroom contributes not only to digital literacy, but also to improvements in mathematics, science, and "learning motivation."

In 2007, the UNESCO Education Council identified 16 core indicators of education and training, many of which directly related to digital literacy and emphasized digital competence.[2] The framework developed stressed the impor-tance of integrating ICT skills into the education system and establishing pro-fessional development for educators through e-learning courses. Subsequent UNESCO reports recognize, however, the difficulties school systems face in developing this integration, and continue to stress the importance of teacher digital competency. Not only must teachers know how to use ICT themselves, but they must also be well versed in methods to utilize ICT to deliver educational instruction.

Beyond acting as a "gateway" for employment, digital literacy skills affect citizens' ability to develop other skillsets. Access to online courses, for instance, can offer both academic and real world instruction, while social and professional networking sites can improve and expand an applicant's job search. As an example, the site LinkedIn now boasts more than 200 million worldwide users, and the majority of its job listings look for future employees in the IT, financial services, and management consulting fields—industries typically offering higher-paying jobs. Without universal digital literacy and the knowledge necessary to navigate through such sites, opportunity stratification will only continue to increase.

3.1.2.1 Adult Education Programs

Adult education programs are focused on upgrading the skills of the workforce, therefore preparing it to fulfill a productive role in the digital economy. They can be structured around conventional continuing education courses, as extension programs of universities, or organized under economic development efforts focused on specific regions of a country.

---

[2] "Digital Literacy in Education." *Policy Brief.* UNESCO Institute for Information Technologies in Education, May 2011. Web. 11 Mar. 2013. http://unesdoc.unesco.org/images/0021/002144/214485e.pdf.

As Hilding-Hamann et al. (2009) concluded in their extensive review of digital literacy programs, a large portion of these programs are targeted to the unemployed, with the objective of increasing their employability. In this context, these programs tend to provide a certification (such as EDCL reviewed above) to provide a proof of skill. On the other hand, digital literacy programs focused on adults with a low education level represent an opportunity to provide a second chance instruction, thereby enhancing their personal development.

Some of the best practices captured in the assessment of adult digital literacy programs include the following:

- Consider delivering courses in mobile settings (e.g. trucks equipped with computers, servers, and mobile broadband) to make it easier for people to participate in different geographies, thus enlarging the reach of the program; the mobile unit and instructors can arrive in one town, install the equipment in a library, a city hall or any community center, offer the five day courses, and then move on to the next location;
- Allow participants to borrow equipment and take it home to continue practicing after the training sessions (although this could face some logistical difficulties);
- The formal course should last approximately five days and be delivered to groups not larger than 12 individuals, so each of them gets proper attention;
- After completion of the formal course, users can enroll in a web-based program; and
- Waive enrollment fee for unemployed adults, but consider charging for others

In this context, certification (proof that the appropriate training was delivered and received) becomes critical. In 2008, Cisco commissioned Forrester Consulting to determine the importance of formal certifications in hiring decisions. By surveying IT hiring managers across the world, the resulting study concluded that "certifications were second only to a college degree to qualify for jobs and the top criteria used in determining ability to perform the job," because they serve to "validate the skills required for computer support technicians and for careers in IT networking."[3]

As an example, the Microsoft IT Academy Program[4] focuses on technology career training. Upon completing the program, students receive certification demonstrating that they have acquired "twenty-first century technology skills." Academy membership offers educators professional development opportunities, technology-centric curriculum and lesson plans, E-learning, student projects, and assessments. At present, Microsoft has more than 10,000 IT Academy members in more than 160 countries.

---

[3] "Courses and Certifications." *Cisco*. N.p., n.d. Web. 06 Mar. 2013. http://www.cisco.com/web/learning/netacad/course_catalog/index.html.

[4] "Microsoft Innovation Center Activities." *Microsoft*. N.p., n.d. Web. 07 Mar. 2013. http://www.microsoft.com/mic/mic-activities.aspx and, "IT Academy Program Overview." *Microsoft IT Academy Program*. N.p., n.d. Web. 07 Mar. 2013. http://www.microsoft.com/en-us/itacademy/overview.aspx.

The program stresses formal certification through the "alignment of academic and vocational standards and courses." The academies provide educators with curriculum mapping to ensure that instruction properly prepares students to receive Microsoft certifications. The mappings pair curriculum and certifications including:

- Microsoft Digital Literacy Curriculum (MDLC)
- Microsoft Office Specialist (version independent and including MCAS)
- Microsoft Technology Associate (MTA)
- Microsoft Technical Certifications
- Microsoft IT Academy learning resources, including E-Learning and MOAC content

---

Case Study 3.12: VOX (Norway)

The Norwegian Ministry of Education and Research promotes workforce competency through its Agency for Lifelong Learning (VOX), which offers adult education beyond basic skill instruction. It focuses on digital competence and ICT skills to improve the age group's employability and educational involvement. While VOX offers a range of courses, it has specialized programs to target groups such as senior citizens, prisoners, and adult immigrants. The agency also conducts research related to adult learning.

Rapid advances in technology and growing involvement in international markets have created demand for a new skillset. Per VOX, more than 400,000 Norwegian adults are "at risk" in terms of their employable skill levels. VOX aims to create a more valuable workforce through basic skill instruction as well as digital and ICT training. The agency administers governmental subsidies and offers financial support. Vox works closely with the European Commission in the development of its adult learning policies.

VOX established a Framework for Basic Skills for adults, emphasizing employable skills and flexibility. One such initiative, "InterAct," promotes on-the-job problem solving through a web-based platform. The activity lasts 5 weeks and targets employees who lack ICT skills. Users log into the website and partake in role-playing activities complete with industry-related scripts that force them to interact and make decisions with other users. Other programs, like "ABC pc," target adults in need of basic ICT training and address tasks ranging from the use of a keyboard to using the Internet and email.
Sources
"Vox in English." N.p., n.d. Web. http://www.vox.no/global-meny/English/.

Government of Norway. *Reporting Template for National Progress Reports in Preparation of the Global Report on Adult Learning and Education (GRALE) and the End of the United Nations Literacy Decade.* Rep. UNESCO, 26 Apr. 2012. Web. http://uil.unesco.org/fileadmin/download/en/national-reports/europe-and-north-america/Norway.pdf.

Case Study 3.13: Technological Specialization Courses (Portugal)

In 1999, Portuguese legislation established technological specialization courses (CETs) to train its adult population and provide them with the vocational qualifications necessary for employment. CETs emphasize scientific and technological knowledge, workplace-relevant skills, occupational placement, and also offer a continuation of studies. Courses typically account for 1,400 h, including 1 year of classroom education plus additional internships or work experience. All students must have a secondary education, although those students in their final year of high school may enroll in CETs.

In addition to the CETs, the government also partnered with industry leaders such as Microsoft, Cisco, and Sun Microsystems to bring ICT Academies to polytechnics and universities across the country. These academies offer students the opportunity to receive professional training from professionals while completing their education. In 2006, three American universities—MIT, Carnegie Mellon, and University of Texas, Austin—partnered with the Portugal Program to add engineering systems, Internet technologies, and digital content to its curriculum.

Portugal also implemented the New Opportunities Program, which targeted adults who did not have a full education. The program offered courses at community centers and local enterprises, all of which involved some degree of ICT skill building. As part of the program, more than 200,000 laptops were distributed.

The country recognized the need for training to address the market for skilled workers. CETs were developed to address Portugal's drop out rates as well as the high proportion of both young workers and mid-level staff lacking the qualifications necessary to succeed in the workforce. The development of the courses focused on the provision of training specific to real-world professional environments.

By 2009, Portugal boasted 119 CETs specializing in ICT-related skillsets offered in 38 institutions across 30 towns. Such courses included multimedia development, information systems installment, and computer programming.

In 2010, the Ministry for Science, Technology, and Higher Education signed a contract with the Portuguese Council of Polytechnic Institutes and several universities, establishing benchmarks to be achieved by 2013. Namely, it aimed to increase the number of placements in CET programs to more than 10,000, with an emphasis on courses offered after working hours.
Sources
Vilhena Nunes Da Costa, Nilza M., Ana R. Simões, Giselia A. Pereira, and Lúcia Pombo. "Technological Specialisation Courses in Portugal: Description and Suggested Improvements." *European Journal of Vocational Training* 46.1 (2009): n. pag. Web. http://www.cedefop.europa.eu/etv/Upload/Information_resources/Bookshop/570/46_en_Costa.pdf.

Magalhães, Luis. "Multi-Program Approach to Foster ESkills." Proc. of WCIT, Amsterdam. UMIC Knowledge Society Agency, 26 May 2010. Web. http://www.umic.pt/images/stories/noticias/Luis_Magalhaes_WCIT.pdf.

"Portugal." European Association of Institutes in Higher Education, Jan. 2011. Web. http://www.eurashe.eu/library/modernising-phe/L5_report_SCHE_in_Europe_addendum_PT_Jan2011.pdf.

3.1.2.2 Digital Literacy for Disadvantaged/Underprivileged Population

Current research points to the fact that one of the largest pockets of broadband demand gap is focused on the disadvantaged/underprivileged segments of the population. However, explanatory variables of this phenomenon are not only economic (addressed below in the affordability section), but also cultural and educational. In this context, any policy oriented to reduce the economic barrier needs to be complemented with digital literacy programs oriented at developing a familiarity with broadband technology.

Given the modern day economic shift away from low-skilled manufacturing jobs to high-skilled services jobs, lack of workplace opportunity is particularly heightened as a result of the digital divide. Additionally, economic, educational, and geographic disparities tend to impact ICT exposure, further exacerbating this cycle. Thus, successful digital literacy programs many times target the following disadvantaged groups that are less likely to have prior knowledge of computers or the Internet and face more hurdles as a result:

- The unemployed
- Older citizens
- Welfare recipients
- Rural populaces

In particular, broadband adoption at home has been proven to be an important contribution to finding employment. Horrigan (2014) found in its survey of low-income families that 62 % explained that they needed broadband to look for or apply for jobs.

Training can be provided in a variety of ways, so long as it is offered in an easily accessible, affordable manner to encourage participation. Many training sessions, for example, are offered at local community access centers or schools, where citizens already feel comfortable, while others are offered online. Sessions can cover a variety of topics, but tend to focus on the development of ICT skills with "real world" application, including, but not limited to:

- Email
- Internet inquiry
- Job search
- CV creation

Many programs also offer certification options, providing participants with tangible evidence of their acquired skillset. Further, as training programs become more popular, they create more economic opportunity through the demand for citizens to serve as trainers or project managers.

Case Study 3.14: IT Mentor Program (Hungary)

In 2003, the Hungarian Ministry of Informatics and Telecommunications established the IT Mentor Program as an answer to its national strategy, which addressed the significant digital divide within the country. The program offered digital literacy training for those disadvantaged members of society who had difficulty entering or re-entering the labor force, such as unemployed and disabled citizens and those citizens over age 45.

After receiving formal certification, the mentors of the program served as social workers, providing training and consultation services as they related to e-knowledge. Such training incorporated basic digital literacy skills as well as awareness; mentors instructed participants not only on basic computer functions but also promoted an understanding of the advantages of such online features as e-Government services and job searches. Beyond addressing the digital divide, the also created "local champions" and strengthened the role of IT trainers in the workforce by offering them government accreditation and support.

The program recognized that these members of society did not have the same access to the digital information and its benefits that could otherwise advance the country's economy as other groups in society. By creating equal opportunities to bridge the digital divide, the program promoted societal and economical development and modernization. As a public policy, the IT Mentor Program received its funding from the government and was managed by the former Ministry of Informatics and Telecommunications.

By 2006, more than 5,000 IT mentors across the country worked with citizens at 20 locations, typically community computer and Internet access points. By encouraging digital literacy training and awareness of the capabilities of the Internet while also providing public access points, Hungary's IT Mentor Program addressed the three pillars of demand stimulation: awareness, affordability, and attractiveness.

As a result of the program, "IT Mentor" became an official profession in the country, allowing employers to search specifically for potential hires with this skill set. In anticipation, an IT Mentor university certification program was established, training 300 mentors in the 1st year. Since then, various government projects—such as the implementation of an online tax filing system—have formally incorporated IT Mentors to promote citizen awareness and serve as troubleshooters.

Sources

OECD E-government Studies: Hungary. Paris: OECD, 2007.

E-Inclusion Public Policies in Europe. Rep. European Commission, 2009. Web. http://ec.europa.eu/information_society/activities/einclusion/library/studies/einclusion_policies_in_europe/index_en.htm.

Case Study 3.15: e-Diriya Program (Sri Lanka)

In November 2011, Sri Lanka's Ministry of Telecommunication and Information Technology and the Hambonthota district inaugurated the country's national ICT initiative, the e-Diriya program. E-Diriya targets Samurdhi (welfare) recipients, many of whom have never before touched a computer. The initiative aims not only to create more than 50,000 computer literate citizens and enhance ICT infrastructure, but also to raise ICT awareness.

The program's 4-h training workshops—which are held in public schools and IT centers across 19 districts—teach basic computer skills to participants with no prior knowledge of computers or the Internet. All centers come equipped with computers and Internet connections. The second-longest segment, which lasts 70 min, covers word processing software, functions, formatting, and outputs. The longest segment, which lasts 90 min, deals strictly with Internet and email, emphasizing the benefits of the Internet, the components required for a connection, and how to utilize a web browser and email address. The other two segments include a basic introduction to ICT and a "getting started" portion that acquaints participants with the hardware, software, and operating system of a computer.

Following completion of the workshop, participants are better prepared to enter the global e-community. As part of the e-Sri Lanka Initiative, e-Diriya uses ICT to "develop the economy of Sri Lanka, reduce poverty, and improve the quality of life of the people." The program aims to reduce the digital divide, thereby creating an equal distribution of opportunity and information.

The Information and Communication Technology Agency (ICTA) manages e-Diriya with the full support of the Sri Lanka Samurdhi Authority and the Ministry of Education. The government fully funds the Samurdhi Program, which covers approximately one-third of Sri Lanka's population, or 1.2 million families. The program encourages poverty reduction by providing disadvantaged groups with training, decision-making activities, and employment opportunities.

Following its November 2011 inception, e-Diriya saw 50,000 participants in December alone, including 23,000 women in the Samurdhi Program. In May 2013, the ICTA opened its fourth e-Diriya center. At this point, the government had provided employment to more than 50,000 graduates of the program.

Sources

Yapa, Seu. "Digital Literacy for 23,000 Rural Sri Lankan Women." *Tele-centre.org Foundation*. N.p., 2 July 2012. Web. http://community.telecentre.org/profiles/blogs/digital-literacy-for-23-000-rural-sri-lankan-women.

"'e-Diriya' the National ICT Literacy Initiative Officially Inaugurated in Kegalle." *ICTA*. N.p., 25 Nov. 2011. Web. http://www.icta.lk/en/icta/90-general/1115-e-diriya-the-national-ict-literacy-initiative-to-be-officially-launched-in-kegalle-tomorrow.html.

"ESri Lanka." ICTA, n.d. Web. http://www.icta.lk/en/e-sri-lanka.html.

"Sri Lanka Case Study Samurdhi Program." *Community-based Food and Nutrition Programmes*. Food and Agriculture Organization of the United Nations, 2003. Web. http://www.fao.org/docrep/006/Y5030E/y5030e17.htm.

"ICTA's 'e-diriya' IT Centre Opens in Kegalle." *DailyFT*. N.p., 21 May 2013. Web. http://www.ft.lk/2013/05/21/ictas-e-diriya-it-centre-opens-in-kegalle/.

Case Study 3.16: e-Ciudadano (Colombia)

In 2008, 200,000 Colombians completed the ECDL Foundation's global computer literacy survey, which demonstrated the social and economic divide between Colombians with ICT access and understanding and those citizens who lacked exposure. In response, the Colombian Ministry for Information and Communications Technology (MITIC) and the National Learning Service (SENA) partnered with ICDL Colombia to increase digital literacy and encourage use of online services such as e-government and banking amongst marginalized sectors of the population, which led to the creation of the e-Ciudadano, or e-Citizen, project. The project expected to provide 75,000 Colombians with basic ICT training and fell under the Colombian national ICT plan *Vive Digital*, which aims to connect all citizens to the Internet by 2019.

Following the training, participants took a certification exam. E-Ciudadano established 88 testing centers throughout Bogota's two main urban areas and offered training at facilities such as public libraries and community centers. The program also featured e-learning courses so that citizens could also participate online.

Implementation of the project was divided into three parts. The first two parts focused heavily on marketing e-Ciudadano; part one addressed participant recruitment while the second worked toward the utilization pre-existing public and private infrastructure to deliver the training. The website www.e-ciudadano.org.com served as an administrative hub. The last phase of the project examined the candidate certification process.

Participants received training at no cost and the first 25,000 candidates took the certification exam free of charge. Fundación Telefónica (Telefónica's branch responsible for promoting educational, social, and cultural growth through increased access to ICTs) and the Colombian government covered these costs. The ECDL Foundation and Webscience A.I., an automated test developer, covered the testing centers costs while volunteers conducted the training.

Initial analysis of the e-Ciudadano program found that successful candidates displayed an increased awareness of the benefits of Internet access, with many seeking additional ICT training upon completion. Further, participants demonstrated higher confidence and reported less social exclusion than many of their peers in the same social group.

The project later extended to government officials. In September 2012, for instance, the program won the pan-Latin American FRIDA prize after training more than 110,000 police officers. At the time of the award, more than 70 % of all Colombian police officers had received the e-Citizen certification, allowing them to utilize e-Government services and connect with the public more efficiently.

Sources

"Colombian E-Citizen." ECDL, n.d. Web. http://www.ecdl.org/media/e-ciudadano_Colombia1.pdf.

"Colombian E-Citizen—Bringing Access to Technology and ICT Skills to All Colombians." ECDL Foundation, n.d. Web. http://www.ecdl.org/index.jsp?p=931.

"E-Citizen Programme Is Awarded for Developing the Digital Skills of 110,000 National Police Officers in Colombia." *ECDL Foundation.* N.p., 27 Sept. 2012. Web. http://www.ecdl.org/index.jsp?n=2803&p=932&a=4481#sthash.ijV5lQp3.dpuf.

3.1.2.3 Digital Literacy for Women

Digital divide based on gender differences has been studied in the emerging world with a varying set of evidence about its level of importance. While explanatory variables of this situation tend to be focused on socio-economic and occupational factors, digital literacy programs with a women focus could act as a contributor to addressing some of the gender barriers.

Most of the digital literacy programs targeted to women have as primary objectives, reduce the digital divide, promote social inclusion and improve the employability profile of women. Alternatively, the program can be focused on educating homemakers (married women not in the labor force) under the assumption that, as principal decision maker regarding household finances, the path to broadband adoption in the home is led through women, as seen in the case of Korea's Ten Million People Internet Education project.

According to Hilding-Hamann et al. (2009), there is no consistent content structure of women digital literacy programs: some involve standard computer courses while others entail courses tailored to specific users' needs. This lack of standardization of program content is because it is common to find users in this group that have strong skills in very narrow ICT areas (e.g. social networking, text messaging), while being weak in others (e.g. conducting Internet queries). In fact, it is very common to find situations where the household already has a computer and broadband access, but due to limited digital literacy on the part of women, the technology is only accessible to the children or their father.

In general, the target of digital literacy programs focused on women comprises the unemployed, low income, low schooling, at home with small or ill children, living in settlements, belonging to marginalized ethnic groups, and the elderly. In many cases, the pressure to be included in the digital society is increasing for mothers of children in school since teaching institutions often use broadband for communicating with parents.

The following best practices in this kind of programs have been identified:

- All instructors should be females with experience in teaching computer skills; students appreciate the notion of "women teaching women", addressing not only a skills gap but providing a remedy to unequal opportunities in the workplace;
- Additionally, the instructors could be unemployed women with prior computer experience; as a result, the program could also become a vehicle for reintegrating unemployed women in the workforce;
- Include a mentoring process in the program, which is based on younger peers or attendees to prior sessions;
- Advertise programs in order to promote enrollment at places such as nurseries, schools, playgrounds, and markets;
- Alternatively, kindergartens and schools could become places for recruiting program participants;
- Provide flexibility in course delivery to allow for occasional absences;
- Structure lessons as "learner-centric" rather than "curriculum-centric", building the program around what attendees say they want to learn (e.g. use online search of job opportunities);
- Consider partnering in delivery of the program with associations or non-governmental organizations focused on advancing women welfare and/or enhancing the social inclusion of women by means of technology;
- If focusing on women belonging to a specific ethnic group, tailor the material to be delivered in suitable language, and customize it to the cultural idiosyncrasies of the targeted group; and
- In some cases, it could be very productive to involve the whole family in learning ICT skills in order to motivate mothers to participate

Case Study 3.17: Telecentre Women: Digital Literacy Campaign (Global)

Partnering with the ITU, the Telecentre.org Foundation launched its global Women's Digital Literacy Campaign. The Philippines-based NGO would utilize its 100,000 worldwide telecenters to provide one million disadvantaged women with basic ICT training by the end of 2012. Over 300 organizations and 200,000 individuals powered the telecenters. The program addressed two types of women: (a) the women who will gain employment serving as trainers and managers at the telecenters and (b) the women who lack formal education and literacy and will benefit from basic ICT skills.

According to the United Nations, 60 % of women in developing countries serve as unpaid workers in their family homes. This program asserts that basic digital literacy will connect women to the technological revolution, offering the opportunity to participate in the global economy. Female agricultural workers, for instance, can use new technology to find market information and better gauge the prices they charge for their products and pay for supplies. Women who stay at home with their children can utilize the Internet to become "homepreneurs" or find Internet-based income opportunities. Per ITU Secretary-General Dr. Hamadoun Touré, "With technology now widely recognized as a critical enabler for socio-economic development, this campaign will further reinforce ITU's global efforts to promote the digital inclusion of women, and will be a key element in achieving Millennium Development Goal 3 on gender equality."

The total budget for the campaign is US$ 149 mn. The largest expenditure comes from the US$ 110 mn investment in the development of community ICT centers. Facility rentals will cost an additional US$ 12.5 mn and trainers' salaries will total US$ 25 mn. Project management and the creation of training materials will account for the remaining US$ 1.5 mn.

The ITU and the Telecentre.org Foundation will encourage national governments, private corporations, and international organizations to contribute to the campaign by offering the telecenters such resources as digital curricula in local languages and trainers. The ITU's Telecommunication Development Bureau (BDT) will also supply digital literacy training materials and curricula, much of which was developed for use in community centers and telecenters. The ITU will also offer its distance-learning platform. Trainers receive training through the telecentre.org Foundation network, offered at academies, universities, and other training centers. To ensure quality standards, the Foundation will monitor all training and track progress on a joint ITU-telecenter.org website.

By December 2013, the program had trained more than 920,000 women, though countless additional women have likely benefitted from the campaign. The program is built on the "Train the Trainer" model, which encourages participants to then train their peers.

Sources

"ITU Launches Global Digital Literacy Campaign for Women." *Newsroom*. International Telecommunication Union, 7 Apr. 2011. Web. http://www.itu. int/net/pressoffice/press_releases/2011/08.aspx.

"Telecentre Women: Digital Literacy Campaign." Telecentre.org Foundation, n.d. Web. http://women.telecentre.org/wp-content/uploads/Campaign-Brief.pdf.

Case Study 3.18: eHomemakers (Malaysia)

Between 1998 and 2000, the Malaysian grassroots organization "Mothers for Mothers" held six conferences, all of which featured the stories of successful women "homepreneurs." The events allowed like-minded women to network and pool resources, sharing their experiences with home-based work.

As the demand for additional conferences continued to grow, a small group of volunteers created the "mom4mom.com" website to provide mothers and homemakers with the platform to network and access information without attending the physical conferences. Unfortunately, many women in the country did not have access to computers and Internet. Even amongst those women who did have access, few possessed the technological savvy to access the portal and employ the Internet's benefits productively in their work. Without this access, women faced even fewer employment opportunities than their male counterparts.

In response, Mothers for Mothers submitted a proposal in 2001 for the eHomemakers grassroots community to the Malaysian Ministry of Science, Technology, and Environment. The ministry subsequently awarded the organization with the Demonstrator Application Grant (DAG). With this funding, mom4mom.com evolved into the website ehomemakers.net, which links homeworkers into an e-community and, at the time, employed more than 60 full-time workers at its virtual office.

Among other services, the website offers interactive tools and chat rooms. More than 17,000 members receive its monthly e-newsletters. eHomemakers has evolved into an e-network connecting more than 15,000 Malaysian women that promotes the use of ICT such as mobile phones, the Internet, and the Distributed Work Management Application (DWMA) web-to-hand platform. DWMA relies on ADSL, Internet access, and mobile SMS to connect the women to the larger network. With this technology, the women can then participate in the economy as homeworkers, tele-workers, or business owners. Members of the community also receive—and assist in— the training, mentorship, and counseling that encourages self-sufficiency.

The majority of members is located in urban areas and is in the 30–50 years age range and has grown to include grandmothers, unmarried women, and even some men who work from home.

While the Malaysian government provided the initial funding through the one-year DAG grant, eHomemakers has generated additional revenue through advertisements on the website and consultancy fees and private contributions have come from corporate sponsorships and member donations. As a social enterprise, it supports itself by providing services for a social purpose.

Source

"Welcome to EHomemakers." *EHomemakers*. N.p., n.d. Web. http://www. ehomemakers.net/en/index.php.

Chong, Sheau C., and Audrey Desiderato. "Empowering Women through Home-Based Income-earning Opportunities in Malaysia." *Poverty Reduction That Works: Experience of Scaling up Development Success*. By Paul Steele, Neil Fernando, and Maneka Veddikkara. London: Earthscan, 2008. N. pag. Print.

Case Study 3.19: DigiGirlz Day (United Arab Emirates)

At DigiGirlz Day, a one-day conference held at multiple locations around the world, high school girls interact with Microsoft employees to learn about the technology industry. Girls also receive career planning assistance and information about technology positions and attend Microsoft product demonstrations. In its mission to encourage girls to consider working in the industry, DigiGirls also hosts its High Tech Camps, offering girls a glimpse into the high-tech product development process. Beyond the conferences and camps, the program also offers online courses with instruction on how to build websites with HTML or create podcasts.

In April 2009, Microsoft partnered with the UAE Ministry of Education and the Center for Women and Technology for the Arab Region (CWTAR) to host the Gulf Region's first ever DigiGirlz Day at Dubai Women's College. 200 girls from 25 Dubai high schools attended the event, where they each participated in one of five product workshops: Microsoft Research AutoCollage, Windows Movie Maker, Windows LiveTM, PopFlyTM, blogging, and Microsoft Expression® Web. Many of the girls admitted that they rarely used computers, and that the program raised their awareness of computer-based services and technology-related career opportunities.

The girls all had the option to take the Microsoft Digital Literacy Certificate Test at one of 10 proctored testing stations. Thirty questions covering basic computing skills comprised the exam, and those girls who passed it

received a Microsoft Digital Literacy Certificate. Of the 200 girls in attendance, 50 took the exam but very few scored high enough to receive the
certificate.

Following the success of DigiGirlz Days Dubai, Microsoft planned an
additional five conferences throughout the region. The low scores on the
certification exam signaled a lack of quality digital literacy education, and in
response, Microsoft included testing stations at future DigiGirlz Days events.

Sources

"DigiGirlz Day." Microsoft, n.d. Web. http://www.microsoft.com/en-us/
diversity/programs/digigirlz/digigirlzday.aspx.

Microsoft. Microsoft Learning. *Microsoft Digital Literacy Inspires High
School Girls at DigiGirlz Dubai.* N.p., May 2009. Web. http://www.
microsoft.com/casestudies/Microsoft-Learning/Dubai-DigiGirlz-Day/
Microsoft-Digital-Literacy-Inspires-High-School-Girls-at-DigiGirlz-Dubai/
4000004417.

3.1.2.4 Digital Literacy Programs in Rural Isolated Areas

Programs focused on rural isolated areas represent a particular case of the
examples presented above. As such, they address the complexities of delivering
training in underserved regions of a country. The primary foci of these programs is
bridging the digital divide and enhancing the employability profile of the targeted
population. In this case, the initiatives tend to be large scale and centrally managed
and focus on accessibility. While the central government plays a prominent role in
program management, it is not unusual to find private sector participants or NGOs.

In many countries, the rural population faces multiple disadvantages over its
urban counterparts: higher unemployment rates and fewer employment opportunities, lower literacy rates, and a larger proportion of citizens living below the
poverty line. To make matters more complicated, these regions have been traditionally underserved in terms of ICT and reflect lower broadband penetration rates.
At the same time, broadband connections could potentially reduce these disparities
and the digital divide. Unlike voice services, it has become increasingly less
expensive to communicate and stay in touch over broadband as well.

Connectivity offers access to services that may otherwise be inaccessible due to
geographical limitations, opening the opportunity to such applications as e-health, e-
government, and e-education. Further, it allows residents to stay connected with friends
and family and informed of current events. Connectivity also means more employment
opportunity, offering the means for citizens to search for and apply to jobs.

Without a perfunctory knowledge of how to use computers and broadband,
however, the advantages of these tools are lost. While government involvement
may spur the development of infrastructure in many instances—particularly as the

private sector sees less financial incentive to invest in areas with more geographical barriers and less disposable income—successful rural training programs many times require assistance from corporations or organizations. The corporations can offer financial support and the benefit of experience, while grassroots organizations may have more success in engaging the people.

As many rural communities may not "be commercially viable on their own," broadband access can afford the utilization of a variety of tools to promote development and sustainability. Rural farmers, for instance, can access weather reports and market information while utilizing agriculture software to improve their business production. Similar examples have been seen for fishermen, basket weavers, health care workers, and the like.

Access is key, as this population cannot easily utilize the physical resources found in urban areas. To this degree, successful programs have:

- Offered online training
- Built local access centers or cybercafés in areas with limited ICT
- Implemented initiatives in public schools or safe houses
- Partnered with local governments
- Deployed trainers to rural areas

Rural broadband training programs bridge the digital divide, ultimately reducing the disparities between the urban-rural populations.

Case Study 3.20: Intel Easy Steps Program (Philippines)

The Intel Easy Steps Program offers basic digital literacy training to adults with very little computer exposure. Topics focus on everyday computer use, from Internet searches to word processing to email. As a result, participants leave the program with the necessary skillset to communicate with friends and family and also to research and apply for jobs, draft resumes, and create presentations. With an increasing number of positions requiring ICT skills, this knowledge increases citizens' employability and self-sufficiency while enhancing the country's economic competitiveness in the international market.

Intel provides the program free of charge to governments and NGOs, which then implement it on a local basis. In November 2010, at the 6th Knowledge Exchange Conference on Community eCenters in the Philippines, the country's Commission on Information and Communications Technology (CICT) signed a memorandum of understanding with Intel Microelectronics Philippines to deploy the initiative. This agreement followed Intel's push to address the lack of digital literacy of the region's government employees and adults in rural communities.

Instruction is offered in a variety of environments, from vocational training centers, to shared-access centers, to the workplace. Intel identified

1,000 community eCenters across the country to host the training and the program considered additional locations for new centers. The small countryside town of Tanuan, for instance, partnered with Intel to create its Community e-Center (CeC), using the Easy Steps program to bring basic digital literacy instruction to its citizens. The manager of the center reported that, since the program's inception, the town saw a rise in the number digitally literate farmers, fishermen, and health workers.

For those citizens unable to attend a workshop in person, Intel Philippines now offers practical digital literacy training through 10 Intel Easy Steps Facebook applications. Each application simulates a real world situation—such as the behavior of a word processor or spreadsheet—allowing users to create such products as resumes and budget plans via the social media website. They can then post their progress as a status message to track accomplishments and share with friends.

Sources

"Intel Easy Steps Program." *Intel*. N.p., n.d. Web. http://www3.intel.com/cd/corporate/csr/apac/eng/inclusion/steps/466382.htm.

"News From the Field." *Intel Easy Steps Program*. FIT-ED, 2012. Web. http://www.fit-ed.org/easysteps/news.php?type=1.

Case Study 3.21: Grameenphone Community Information Centers (Bangladesh)

In late 2009, Grameenphone—Bangladesh's largest mobile operator—partnered with Microsoft to implement a digital literacy program developed specifically for the country's rural students. As a joint venture between international telecommunications provider Telenor and the Grameen Telecom Corporation, which is affiliated with the micro-finance pioneer Grameen Bank, Grameenphone has a history of promoting affordable telecom services throughout the country.

This partnership took Microsoft's well-established international digital literacy curriculum and first reproduced it in Bengali and then targeted the country's rural students, unemployed youth, and women. The module focuses on basic IT skills such as Internet access, applications to increase workplace productivity, and computer security. Beyond coursework and training, the program also offers related resources for further learning and self-assessment exercises. Each section concludes with an examination, and students who successfully pass receive a certificate.

Citizens can access the curriculum through other Grameenphone projects such as school cyclone shelters, Information Boats, and education institutions as well as through the more than 500 authorized Grameenphone Community

Information Centers (GPCICs). Since 2006, GPCICs have provided the rural population of Bangladesh with access to Internet, voice, and video conferencing services. Each center houses at least a computer, a printer, a scanner, a web cam, and an EDGE-enabled modem that allows for Internet connections. While the centers receive technical support from the GSM Association, local entrepreneurs manage and run them like small businesses. Grameenphone provides these managers with training and support.

The vast majority of the country lives in rural areas; by targeting this sector and providing them with ICT access, the program effectively reduces the geographical digital divide. In turn, the GPCICs aim to alleviate poverty, bring education to the underprivileged, create employment opportunities for the youth, and promote local entrepreneurship. Grameen has worked with many international corporations, organizations, and multilaterals in support of this mission. Microsoft provides citizens with the digital literacy curriculum and the examination free of charge. In the partnership, Grameen is the licensee, Microsoft the licensor, and the GPCICs and other Grameen centers the authorized centers.

Sources

"Grameenphone—Microsoft Digital Literacy Programme." *Telenor*. N.p., 16 Apr. 2012. Web. http://telenor.com/corporate-responsibility/initiatives-worldwide/grameenphone-microsoft-digital-literacy-programme/.

"About Us." *GPCIC: Grameenphone Community Information Center*. N.p., 2007. Web. http://www.gpcic.org/index.php?main=0.

"Grameenphone Partners with Microsoft." *Grameenphone*. N.p., 4 Nov. 2009. Web. http://www.grameenphone.com/about-us/media-center/press-release/2009/239/about-us/career.

Case Study 3.22: Ford-Rotary Digital Literacy Program (India)

In 2010, Tamil Nadu's Ministry of Information Technology commenced its digital literacy program targeting the state's rural population. The program—a partnership between Ford Business Services and Rotary International—provides beneficiary institutions with the hardware and software necessary to provide digital literacy training.

Through the alliance, Ford Business Services—a subsidiary of US-owned Ford Motor Company—initially donated 150 computers to 11 partner groups. The groups included organizations that focused on women empowerment, underprivileged youth, and rural communities as well as vocational training institutions. Computer novices can now access the 20-h training online through any of the participating institutions. The curriculum includes six modules and uses the digital literacy CD "Know It," designed to train citizens in basic computer functions, such as the use of the Microsoft suite, Internet, and email. Users interact with the CD's step-by-step online teaching tool that also offers practice quizzes and activities. All software was locally developed with rural users in mind.

The State of Tamil Nadu has demonstrated a commitment to enhancing ICT access and understanding throughout the region, recognizing the potential for this technology to bridge the digital divide. Prior to this initiative, it had already created an IT plan and an e-waste policy while investigating ways to increase e-government services. While Tamil Nadu is one of India's most urbanized states, its rural village population is marked by lower levels of literacy and fewer employment opportunities, both of which affect economic development in the region.

Since the program's inception, Ford has donated more than 700 computers and connected nearly 10,000 people in the state. In September 2012, Ford India inaugurated the Ford-Rotary Digital Literacy Center. The center provides the underprivileged population with the computer literacy training necessary to enhance their employment options. The center offers multiple courses throughout the day, instructing participants on basic computer use.

Sources

"Digital Literacy Programme to Empower Rural Population." *The Hindu*. N.p., 17 Aug. 2010. Web. http://www.hindu.com/2010/08/17/stories/2010 081754270400.htm.

"Ford-Rotary Partnership Helps Bridge Digital Divide in India." *@FordOnline*. N.p., 8 Oct. 2012. Web. http://www.at.ford.com/news/ TeamContent/Pages/GFTCFord-Rotary-Partnership-Helps-Bridge-Digital-Divide-in-India.aspx.

India. Tamil Nadu. *11. Rural Development*. N.p., n.d. Web. http://www. tn.gov.in/dear/11.%20Rural%20Development.pdf.

3.1.2.5 Digital Literacy for Persons with Disabilities

Recent literature has highlighted the need to deploy digital literacy programs targeted to persons with disabilities (e.g. visually impaired, hearing impaired, etc.). Most of these programs respond to the need of improving the quality of life and promoting social inclusion of the disabled population. Their focus is to build broadband and Internet usage capabilities. Courses are tailored to the specific needs of the disadvantaged population.

While most of these programs are implemented at the national scale, it is not uncommon to observe the participation of non-government organizations (e.g. interest groups, community organizations) as program stakeholders. Their participation is structured around partnerships with public agencies. The advantage of including these types of organizations in the structuring and delivery of digital literacy programs is that they tend to have a better understanding of the needs of the group being targeted.

While most programs of this type have been deployed in the industrialized world (Hilding-Hamann et al. (2009) surveyed 95 such programs in their review of EU initiatives), the examples provided below are used to demonstrate the benefits of extending these initiatives in emerging countries.

Among the best practices in the deployment of digital literacy for persons with disabilities, the following have been highlighted by the authors:

- Organize the digital literacy programs in centers already focused in helping disabled citizens;
- If physical facilities focused on the disabled do not pre-exist, ensure that new centers are handicap friendly, equipped with special tables, with plenty of room allowed for wheel chairs. Computer should also be designed to fit the needs of the disabled;
- From an equipment perspective, special customized hardware should include voice output devices, special large-surface keyboards, trackballs, voice activated devices, etc.;
- The centers and digital literacy training should be opened to people with all kinds of disabilities: physical, psychosocial, hearing impaired, cognitive difficulties like dyspraxia, dyslexia, dysorthography, intellectual disabilities, etc.;
- Training should be adapted to each disabled situation, resulting in the development of special training materials;
- However, start-up training modules should include internet access, word processing, spreadsheet usage, e-mailing, digital presentations, information search, and web-banking;
- Try to enroll professionals trained to work with people with disabilities (e.g. special needs teachers, occupational therapists, social workers, etc.); these professionals should be trained to deliver ICT literacy programs;
- Be prepared to handle individualized coaching;
- Consider distributing used computers donated by private companies that participants could take for use in their homes and workplaces;
- While computers are provided free of charge, consider charging a nominal fee for course attendance to create an incentive for ongoing participation;
- Conduct on-going monitoring of the latest development of technologies to support people with disabilities;
- Structure a feedback mechanism, where participants fill out feedback forms to improve training and adapt it to the needs of attendees; this could be complemented with random follow-up calls to former participants;
- Consider organizing a placement function to help participants find jobs; the placement function could organize a portal to link up with companies interesting in providing employment opportunities to disabled individuals; and
- Funding could be organized with contributions from the private sector (a telecommunications company provides broadband access, a software company provides the training modules, one-time contribution for hardware acquisition)

and the educational system (universities could supply technical support and *pro bono* training staff); this could allow programs to be set up without any dependence upon government funding.

Case Study 3.23: DICOMP-S.net Project (Europe)

The Digital Competence Screenreader Network—or DICOMP.S-net—project ran from January 2007 through June 2008, focusing on increasing digital literacy within Europe's blind and visually impaired population.

The project provided a screen reader designed for blind and visually impaired participants free of charge. The screen reader was compatible with the Microsoft suite to promote the utilization of technology applicable to every day life. To offer ICT use instruction and support, the program recruited blind and visually impaired tutors and provided them with training so that they could in turn assist participants with implementation of the screen reader and demonstrate its functions. Tutors were trained using an e-learning application.

The product dissemination phase began by distributing the screen readers to partner groups in supporting countries, which then conducted pilots and testing to suggest necessary adaptations for their specific populations. All content, such as texts and tools, was translated into the most relevant language of the country so as to reach and appeal to the largest possible group.

The European Commission funded the project in part and partners included training institutions and non-profit organizations across Europe. Berufsförderungsinstitut Steiermark (Vocational Promotion Institute) took ownership of the project. Founded and owned by the Austrian Federation of Trade Unions and the Chamber of Labor, the non-profit has offered vocational training and adult education for more than 50 years and the European Union has commissioned many of its projects. While specific financial data was not released, the approximate implementation costs fell in the €300,000–€499,000 range. The economic impact was valued in the €49,000–€299,000 range.

Per an assessment by the European Union, the DICOMP.S-net program saw the most success through the training of tutors for blind and visually impaired screen reader users, which took place partially through an e-learning application.

Sources

"Digital Competence Screenreader Network." *Dicomp-S.net*. N.p., n.d. Web. http://www.screenreader4free.eu/aims.html.

"Digital COMPetence Screenreader NETwork." *EPractice.eu*. European Union, n.d. Web. http://www.epractice.eu/en/cases/dicompsnet.

Case Study 3.24: National Deaf-Blind Equipment Distribution Program (United States)

In 2010, President Obama signed the twenty-first Century Communication and Video Accessibility Act (CVAA), ensuring that the visually and hearing impaired population had access to communication technology such as "video, voice, text, and other capabilities of smartphones, digital television, and internet-based video programming."

The CVAA addresses such technology as VoIP, Internet streaming, and smartphones. Manufacturers must now also create devices with the needs of the disabled population in mind, incorporating, for instance, means for the visually impaired to view Internet pages and emails. All devices that receive or play video programming must have closed-captioning and video description capabilities and provide access to emergency information.

In 2012, the Federal Communications Commission (FCC)—the agency managing the implementation of the CVAA—included the National Deaf-Blind Equipment Distribution Program (NDBEDP) as part of the project. In partnership with the Helen Keller National Center and the Perkins School for the Blind, the program provides the necessary equipment and training to help the low-income blind-deaf population connect with the community through the use of ICT. Equipment can include hardware, software, and applications so long as it is designed to address telecommunications accessibility.

The FCC began the program's pilot by designating one certified entity in each state to distribute the equipment and, in some cases, provide training on how to use the equipment. The iCanConnect campaign provides outreach, assessments, telecom technology, and training free of charge to individuals who qualify for the NDBEDP.

The CVAA permits the FCC to spend up to US$ 10 mn per year from the Telecommunications Relay Service Fund to support the program. The pilot program allocated a minimum US$ 50,000 to each certified distribution entity with additional funding commensurate with state population size. The Perkins School for the Blind will also receive US$ 500,000 annually to coordinate and promote the program.

The program was slated to run from 2012 through 2014, but will now continue through June 2015.

Sources

Pike, George H. "President Obama Signs the twenty-first Century Communications and Video Accessibility Act." *President Obama Signs the 21st Century Communications and Video Accessibility Act.* Information Today, Inc., 11 Oct. 2010. Web. http://newsbreaks.infotoday.com/NewsBreaks/President-Obama-Signs-the-st-Century-Communications-and-Video-Accessibility-Act-70569.asp.

"Guide." *National Deaf-Blind Equipment Distribution Program*. FCC, n.d. Web. http://www.fcc.gov/guides/national-deaf-blind-equipment-distribution-program.
"ICanConnect." *ICC*. National Deaf-Blind Equipment Distribution Program, n.d. Web. http://www.icanconnect.org/index.php.

3.1.2.6 Digital Literacy for the Elderly

Generational differences represent another major barrier to broadband adoption. As mentioned in Chap. 2, typical age cohort where adoption starts declining dramatically in emerging countries is 30 years old (when controlling for income). In that sense, digital literacy programs conceived as extension of either universities or secondary schools have proven to be very valuable in bridging the generational gap. The overall long-term goal of these programs is to improve social inclusion of the elderly population. The primary content delivered in this type of programs are standard computer courses, in some cases tailored specifically to the needs of the elderly (e.g. email to communicate with the family, photo sharing, use financial applications, purchasing tickets online, etc.). However, in addition, digital literacy courses for the elderly give seniors an opportunity to meet people and develop a social network.

The advantage of including non-government organizations in the case of programs for the handicapped population discussed above is also applicable to initiatives focused on the elderly.

Among the best practices in the deployment of digital literacy for the elderly, the following have been highlighted by Hilding-Hamann et al. (2009):

- Carefully determine needs of targeted population given the different requirements that have been observed across the segment;
- Create a website supporting the program, which would include self-study course modules for use on an ad hoc fashion in community centers;
- Self-study programs should comprise online courses, complemented with traditional printed materials;
- Include an entertainment section (media, music) in the training website to enhance attractiveness;
- Strive to coordinate the program with cultural organizations that are part of the user community (for example, they can act as advertising vehicles for digital literacy programs);
- Equip program with self-contained units that could be used via touch screens and a simple menu system;
- If program is offered at a community center, ensure continuous presence of host instructors that can answer inquiries, take registrations, and be responsible for all technical logistics;

- Make sure that instructors stay after classes to act as tutors for the seniors that stay in the center working on the computers;
- Provide an environment where users can share their experiences in dealing with technical issues with peers, which constitutes an important retention mechanism;
- Digital literacy programs for the elderly attain better results when they are delivered in an environment that provides the opportunity to meet other people and break their social isolation;
- It is sometimes useful to involve students of upper secondary schools in the role of volunteer "digital facilitators" to teach internet browsing and e-mail use to the elders; the one-to-one relationship between the young tutor and the trainee (a concept called "intergenerational learning") improves the learning experience; and
- Focus on teaching material that is immediately transferable and applicable to the senior everyday life.

Case Study 3.25: Senior-Info-Mobil (Germany)

In 1996, the German federal government established a task force to promote inclusion of senior citizens within the growing information society. Under this task force emerged the "Older People in the Information Society" working group, which brought to light senior citizens' lack of exposure to ICT otherwise found in an educational or professional context. The committee proposed the Senior-Info-Mobil project, which brought technology to the people by way of a mobile Internet café that offered exhibits and training to raise awareness of ICT's capabilities.

Most of the project's tutors were also senior citizens and the group-oriented instruction created a more comfortable environment in which to learn about and adapt to the new technology. Many of the participants reported that prior to Senior-Info-Mobil, they had never before used a computer or the Internet. A study revealed that, following the project, 50 % of participants had interest in learning more about the Internet. 70 % of visitors to the bus were over age 60, and more than 20 % were at least 70 years old.

The Federal Ministry of Economic Affairs provided financial support for Senior-Info-Mobil. Partners IBM Germany and German Telecom donated the two-level omnibus and telephone connections, respectively, while T-Online sponsored the Internet accounts. The bus was re-designed and equipped with a Local Area Network (LAN) and six Internet terminals. It also featured a PC designed specially for blind and visually impaired people. Additional PC and AV equipment allowed tutors to host presentations.

Private sponsors funded and managed promotion campaigns, public relations, and volunteer coordination. Through media attention and advertisements, more than two million citizens over age 50 heard of the program through television or radio and nearby European countries requested that the bus travel to their countries.

In the first 3 years of the project, the mobile café traveled to more than 60 German municipalities, with over 60,000 citizens between the ages of 50 and 96 visiting the Senior-Info-Mobile bus and taking part in trainings and demonstrations. Beyond directly impacting participants who visited the bus and took part in the demonstrations and training, the project inspired organizations to incorporate senior ICT training into their work plans.

Sources

Senior-Info-Mobil: A German Awareness Rising Campaign on IST Targeting Older Citizens. Rep. SeniorWatch, 28 Dec. 2001. Web. http://www. seniorwatch.de/cases/01.pdf.

Case Study 3.26: SeniorWeb (Netherlands)

In the 1990s, the Netherlands' Ministry of Economic Affairs partnered with the Ministry of Health, Welfare, and Sport to finance a program that would promote ICT use amongst its older population. Along with other suggestions on how to address this goal while at the same time providing digital literacy training came the SeniorWeb portal, which offered activities and campaigns designed with senior citizens in mind.

Launched in 1996, the website *seniorweb.nl* now serves as a home page for citizens in the Netherlands over age 50, offering information, advice, and links to Internet-related websites. It emphasizes user interaction, encouraging members to place advertisements or take part in activities like photo sharing contests. On their own, users have developed chat rooms, discussion lists, and "web families." Seniors can customize their individual experiences by choosing to utilize different features ranging from the helpdesk to email listservs that connect members with similar hobbies and online courses.

Membership includes digital literacy training courses conducted by SeniorWeb "ambassadors." The ambassadors are all volunteer senior citizens who understand the needs of this age group. They not only teach the lessons, but also organize the courses and make arrangements with local educational institutions or Internet cafes while offering technical support as needed. The face-to-face lessons are especially valuable for those senior citizens with low education and literacy levels who may otherwise not benefit from on-screen instruction requiring considerable reading comprehension skills.

The ultimate goal of the program is to promote online participation amongst senior citizens. The project recognizes that the older population may require additional training to increase ICT awareness amongst the older population, which did not grow up using computers or benefit from digital literacy courses in the education system.

To join SeniorWeb, each member pays an annual €28 (US$ 38) fee. The fee includes access to all areas of the website as well as a subscription to the quarterly magazine, *Enter*, and an annual CD-Rom. Partner organizations offer discounts on their products to members.

SeniorWeb's network now includes 3,000 active ambassadors and 145,000 members and the site sees more than 300,000 visitors per day. Research and analysis of the program indicates that the program has increased digital literacy amongst an even higher share of the population as members informally share their learned knowledge with their families, friends, and members of the community. The program has expanded to reach the 50 + population in Germany, Austria, and Switzerland.

Sources

"SeniorWeb." *SENIORWEB*. N.p., n.d. Web. http://www.seniorweb.nl/english.

"EInclusion Factsheet—Netherlands." *EPractice.eu*. European Union, 4 Jan. 2011. Web. http://www.epractice.eu/en/document/5265691.

"Crossing the Bridge between Functional and Digital Literacy." N.p., n.d. Web. http://www.uni-kassel.de/~ifriedri/Crossing_the_Bridge_NL.pdf.

Case Study 3.27: Senior Connects (United States)

Senior Connects began offering digital literacy services to senior citizens in 2003, increasing their computer access and ICT skills by working with senior centers, retirement apartments, and independent living facilities. High school student-volunteers teach basic computer and Internet courses and in some instances, their schools provide the computers and broadband access necessary for the training. The program is based on the "train the trainer" premise; volunteers first receive training before working with participants.

Participants do not pay to take the classes, which typically only have 1–2 students per tutor. Using the pre-established Senior Connects Methodology, the program is customized for participants who may have little or no experience with computers.

Starting a Senior Connects program at a living facility, senior center, or senior apartment complex does not require any cost. Ideally, the chosen facilities will already have computers and Internet access, but Senior Connects offers resources for individuals or organizations looking for funding. Senior Connects operates as a 501(c)(3) not-for-profit corporation and accepts donations.

The Senior Connects website offers instructions for potential volunteers, with topics ranging from establishing contact with senior centers to interacting with students to conducting the actual training. It also provides detailed surveys and lesson plans that the volunteers can use as well as information on college scholarships, whitepapers, and useful resources.

In the first year alone, the Senior Connects program worked with more than 11,000 residents and this number exceeded 50,000 in 2010. By 2008, the program had provided computers to more than 100 senior citizen facilities. Senior Connects now operates as part of the Net Literacy project, a larger initiative that came to fruition as Senior Connects offered more services to other sectors of the population.

A study of the program demonstrated the achievements of Senior Connects. 80 % of participants had never before used a computer prior to the program, but one year following the course, 93—96 % of senior students reported using the Internet more than twice a month. The study attributed the following factors, amongst others, to the Senior Connects' success: one-on-one training, training provided free of charge, building computer labs where seniors already lived, modifiable training materials, and emphasizing the value of Internet use in connecting with friends and family.

Sources

"What Is Senior Connects?" *Senior Connects.* Net Literacy, n.d. Web. http://www.netliteracy.org/senior-connects/.

3.2 Community Access Centers

Deployment of community access centers can assume different forms. In some cases, broadband access centers are installed in existing facilities as a complement of activities already performed at those locations (e.g. public libraries). Other practices point to the creation of stand-alone access centers exclusively focused in providing free access to broadband services.

This section will review the different options, paid or free, for providing broadband access. The objective is to enhance Internet access in contexts where affordability represents an insurmountable barrier to adoption.

3.2.1 Types of Shared or Community Access Centers

3.2.1.1 Digital Community Centers

Digital community centers represent the most common approach to providing public access to broadband, while organizing technology awareness and education programs. The deployment of community centers combines a top-down and bottom-up governance framework, whereby a public policy initiative triggers the involvement of communities in the management of each unit. The sum of grass-root community organizations dedicated to managing each center is coordinated by a steering committee, who works with each center to develop plans for extending broadband service, and providing technology awareness and training programs. In some cases, the steering committee works with a dedicated staff that acts as a resource. In that sense, the central dedicated staff becomes an enabler of the community-based effort rather than an implementer.

By virtue of their decentralized governance framework, centers become independent from contributions of the national government, with all funding support being provided by either local governments or the private sector. This structure appears to be also scalable across regions of a given country.

Digital Community Centers have become highly suited to tackle technology and economic development programs within rural contexts.

The following best practices have been identified in this domain by Hilding-Hamann et al. (2009):

- Establish a permanent channel of communication between the community and the managers of the community access center, involving the community directly and encouraging to take ownership of the activities of the center;
- Community involvement could entail nominating local technology champions, who assemble community support, lead technology needs assessment and planning efforts, and work to introduce technology initiatives to meet community needs;
- Construct digital community centers as a technology and entrepreneurship hubs within communities; as such, the centers provide free broadband access to the public, and, at the same time, a variety of fee-based business and technology services to local non-profits and businesses;
- Among the entrepreneurship services that the digital centers can provide are employee training, modern office space, technology expertise and business consulting;
- Put in place a full technical service team that ensures that all equipment is always working properly;
- Besides hosting IT courses and provide access, make sure that the center functions as venues where people can meet and enjoy other cultural activities and entertainment;
- In terms of advertising and promotional activities, the center should post monthly newsletters on its website, addressing issues for small businesses, such as fundraising opportunities, or dealing with foreign worker authorization permits; and
- Consider outsourcing some of the center functions to facilitate its sustainability.

Case Study 3.28: Pontos de Cultura (Brazil)

As part of the Brazilian Ministry of Culture's larger program, Cultura Viva, the *Pontos de Cultura* initiative is a socio-digital inclusion program that develops public digital spaces throughout the country to encourage citizens to create digital culture. By providing citizens with free, open-source software and broadband access at these telecenters, the initiative promotes technology as a tool to spur the spread and creation of digital culture, thereby affirming Brazil's cultural identity.

Individual communities take charge of their ponto's financial matters, managing the center autonomously, although they all have access to a network over which they can work together to share ideas and problem solve. The pontos have the potential to generate income for these communities, which can customize the services of the centers to fit the needs of their residents. Once the Ministry of Culture deems it an official Ponto de Cultura, the center receives a digital multimedia kit, which guarantees users broadband access so that they can share their work. It also includes a multimedia studio complete with professional-grade audio, video, software development, text, and imaging technology. Equipe Cultura Digital and local grassroots organizations offer training on how to use these tools and also on the benefits of broadband in transmitting their ideas.

As part of the program, the Cultura Digital Equipe (digital culture team) hosts workshops that focus on educating the community on how to use new technology to best suit their needs. The pontos de cultura receive a monthly stipend of €1000 for the first two years, at which point they should sustain themselves. The GESAC Program of the Ministry of Communications provides this funding as part of the aforementioned media kit.

While the centers are run autonomously and funding lasts for two years, the pontos must continually report their progress to the Ministry of Culture to ensure that they stay on track and align with the Ministry's overall mission of promoting digital culture. The program has faced criticism for not establishing qualitative indicators by which to judge its efficacy. There are currently 22,500 pontos de cultura throughout the country.

Sources

Bria, Francesca, and Oriana Persica. "Synergies between Pontos De Cultura and Ecosystems." *Digital Ecosystems*. By Matilde Ferraro. N.p.: n.p., n.d. 4.5.1-.5.8. Web. http://www.digital-ecosystems.org/book/pdf/4.6.pdf.

"Ministério Da Cultura." *Study Tour Brazil*. N.p., n.d. Web. http://studytourbrazil.wordpress.com/rio-de-janeiro-2/ministerio-da-cultura/.

"Pontos De Cultura." *Cultural Exchange Brazil*. Dutch Culture, n.d. Web. http://www.culturalexchange-br.nl/organisations/pontos-de-cultura.

Case Study 3.29: Puntos de Acceso (Venezuela)

Venezuela's Puntos de Acceso program marked the country's first step toward the provision of universal telecom services. The program aims to bring fixed-line and Internet services as well as proper training to the general public, which will allow citizens to use ICT effectively and increase productivity and improve their quality of life. Each access point features ICT services and training that focus on educational, cultural, and economic activities while encouraging information exchange and increased communication. Each center also makes provisions for the people with handicaps or special needs.

Per CONATEL, Venezuela's telecommunications regulatory authority, the sectors of the population who do not have access to ICT services also tend to suffer from social exclusion, economic insufficiency, and a lack of basic essentials like food, shelter, and education. Puntos de Acceso can provide telecom access to these groups Workshops and seminars are also available to teachers and professors, who can in turn share this instruction with their students.

Prior to implementation, the project first examined the demographics of the country as well as its existing ICT infrastructure, electricity supply, and potential demand. The Puntos de Acceso initiative then targeted four states within the country based on these factors, strategically choosing locations equidistant from areas void of ICT services.

The proposal for the project does not list costs specifically, as the cost of the infrastructure and operation vary by operator. Per the Telecommunication Statutory Law (LOTEL), the country's Universal Service Fund will subsidize the project's infrastructure expenditures while promoting market competition. The law requires telecom operators doing business in the country to contribute 1 % of their gross income to the fund in order to finances these costs. The fund, established in 2000, serves to "offer minimum penetration, access, quality, and economic accessibility standards, regardless of geographical location" throughout Venezuela. CONATEL sets the standards for quality and technical requirements.

To fulfill the obligations of the Universal Service project, Puntos de Acceso addresses: international standards, functionality, scalability, adaptability, and facility of administration. Each punto must provide a two-way traffic IP to assure accessibility. For the purpose of the project, CONATEL made available the following frequency bands: 2300–2400 MHz, 5725–5850 MHz, 10,27–10,30 GHz, and 10,62–10,65 GHz. The program also makes provisions for a Social Management Program (SMP) to serve as a methodological tool to stimulate new projects and oversee qualitative evaluation.

Sources

Puntos De Acceso: First Telecommunications Universal Services Obligation in Venezuela. Rep. CONATEL, n.d. Web. http://www.itu.int/ITU-D/fg7/case_library/case_study_2/Americas/Venezuela_English_.pdf.

"Puntos De Acceso." CONATEL, n.d. Web. http://www.conatel.gob.ve/index.php/principal/puntosdeacceso.

Case Study 3.30: Argentina Conectada (Argentina)

In October 2010, Argentina's president Cristina Fernandez de Kirchner announced a five-year plan—and an initial investment of US\$ 1.7 bn—for its "Plan Nacional de Telecomunicacion," also known as "Argentina Conectada." Backed by public investment, the plan covers the deployment of necessary equipment, infrastructure, and services to provide ICT equipment and training to public school students, extend connectivity to remote areas, and establish public access ICT centers. Through this initiative, the government dictates the promotion of broadband growth and the equitable distribution of such services.

The first phase of Argentina Conectada deployed 28 knowledge access centers—known as Nucleos de Acceso al Conocimiento, or NACs—in public areas throughout the country. Each NAC provides ICT access and training as well as entertainment and cultural applications. The training increases skill development while promoting engagement in community affairs. Through the NACs, users have access to WiFi Internet, personal computers, audio-visual equipment, and gaming consoles. The plan developed the NACs not only to provide Internet access, but also to promote e-inclusion and community participation. They offer educational tools and technology and serve as a point of contact between citizens and their government.

Once this first phase of Argentina Conectada reached completion, the plan initially proposed a total of 250 NACs, although the project experienced unexpected levels of success. In fact, in 2014, Argentina Conectada announced a proposal to develop a total of 250 NACs.
Sources
"Argentina Government Deploys 28 Knowledge Access Centers." *Telecompaper*. N.p., 2 Aug. 2012. Web. http://www.telecompaper.com/news/argentina-govt-deploys-28-knowledge-access-centers.

"Nucleos De Acceso Al Conocimiento." *Argentina Conectada*. N.p., n.d. Web. http://www.argentinaconectada.gob.ar/contenidos/nucleos_de_acceso_al_conocimiento.html.

"El Ministerio De Planificacion Sumara 250 "NAC" En Todo El Pais." *El Comercial*. N.p., 22 Feb. 2014. Web. http://www.elcomercial.com.ar/index.php?option=com_telam&view=deauno&idnota=417751&Itemid=116.

Case Study 3.31: Thaicom Digital Communities (Thailand)

In 2009, the satellite provider Thaicom launched its Kids Thaicom Project, donating satellite dish televisions throughout schools in Thailand's remote regions. Deemed an example of "innovative corporate social responsibility," the program not only focuses on the provision of the equipment, but on fostering an understanding of the benefits of technology amongst the country's students by connecting them with worldwide mentors, teachers, and leaders. As the president of the corporation explained to the children, "They are tools for you too. Not just city kids with money."

Prior to the project, Thailand identified 10,000 rural schools with an inadequate teacher: student ratio. By virtue of their location, these schools did not have the means to benefit from the technological advances of the twenty-first century and its resources. Based on this criterion, Thaicom then selected 999 schools to participate in the project.

The schools each received a Thaicom satellite dish, DTV receiver, and TV, installed by both providers hired by Thaicom and by the corporation's more than 300 volunteers. Thaicom also covered the connection costs and offered ongoing training and technical support as needed. Once connected, the schools could access the Internet and educational programs, while having the freedom to incorporate the technology into the curriculum as they saw fit.

Within a year and a half of the project's implementation, the project had connected more than 759 schools. By mid-2011, Thaicom expanded the program to use the connected schools as "sustainable knowledge centers," to develop vocational training and small businesses in rural areas. Thus, providing the schools with the technology not only improved the students' education, but also raised community awareness. The involvement of local volunteers and Thaicom employees is credited with the program's success, creating a sense of pride amongst the community.

As an extension of the project, in 2012 Thaicom produced more than 76 h of specialized satellite television programming designed to prepare Grade 6-level students for their standardized exams. The programs aired on weekdays from October 2012—January 2013, with reruns broadcast on weekends.

Sources

Zacharilla, Louis. "The Future of Thailand—999 Schools beneath the Enlightened Footprint." *Digital Communities*. N.p., 25 Mar. 2011. Web. 19 Nov. 2012. http://www.digitalcommunities.com/blogs/communities/The-Future-of-Thailand—999-Schools-beneath-the-Enlightened-Footprint.html.

"Corporate Social Responsibility." *Thaicom*. N.p., n.d. Web. http://www.thaicom.net/about_csr_en.asp.

Case Study 3.32: e-Mexico National System (Mexico)

In 2003, under the direction of President Vicente Fox, Mexico launched its "e-Mexico" initiative, developing 10,000 "Digital Community Centers" to connect 90 % of the country to the Internet and increase the number of Internet users within the country from 4.5 to 60 million users. The project was developed to reduce the country's digital divide, making Internet access universal regardless of location or socio-economic factors.

E-Mexico focused on three overarching themes: connectivity, content, and systems. It not only offered the provision of physical centers where citizens could access the Internet, but also promoted its application through e-government, e-health, e-economy, and e-learning services. It also encouraged knowledge sharing through increased connectivity. Nearly three-quarters of all center activity was based on some form of digital literacy training or learning service.

The project did not have a specified budget, but rather focused on developing partnerships with various sectors to achieve the most cost-efficient deployment as possible. Mexico's ministries and foundations collaborating on this project included the Public Education Ministry, the Latin American Institute of Educational Computing, the Health Ministry, the National Federalism Institute, the Education for Life and Job National Council, and Adults Education National Institute.

Within a matter of months, the project had completed the installation of more than 3,000 sites, mainly within schools, libraries, medical clinics, and post offices. The second phase of the project installed an additional 4000 network access points, and by the end of 2004, the project had brought a total of 7200 CCDs to the country. E-Mexico reached its goal of 10,000 CCDs by 2006. Its interactive portal was featured at 7500 sites and saw more than 5 million monthly visitors.

Sources

"Intelligent Community Visionary of the Year." *Intelligent Community Forum (ICF)*. N.p., 2012. Web. https://www.intelligentcommunity.org/index.php?src=gendocs.

"E-México: Mexico's Digital Community Centers." N.p., n.d. Web. http://bos.fkip.uns.ac.id/pub/onno/library-ref-eng/cd-apec-telecenter/docu ment/29_15_e-Mexico_CCDs_JPM.pdf.

"E-Mexico National System." Proc. of Wireless Internet Society North America Digital Cities Convention, Houston. N.p., 2 Mar. 2006. Web. http:// broadbandadoptiontoolkit.com/download/p/fileId_81.

Case Study 3.33: Hewlett-Packard Digital Community Centers (Hungary)

Over the course of four years, Hewlett-Packard implemented Digital Community Centers in twelve countries throughout the Europe and MENA region as part of its commitment to provide underserved communities around the world with the necessary tools and infrastructure for learning and development. The centers provide low-income communities with basic computer, Internet, and business skills to help its citizens find employment. The project typically targets teachers, health care providers, students, and the unemployed. Participating communities were selected based on the capacity of ICT to further their development as well as "their capacity to execute and sustain their vision and plans." Hewlett Packard's presence in the region also contributed to the selection process, as did its relationship with local organizations and businesses, many of which contributed to the project.

December 2002 marked the launch of the Hungarian Digital Center in the city of Miskolc at the country's Petroleum and Gas Institute. The DCC offered the country's first multimedia center of its kind to train teachers and students on how to implement online training within the environmental industry. In 2005, the center launched its "Envirotrainer" program, a distance-learning program designed to train secondary school teachers on the implementation of environmental technology into the classroom. Program topics include water and waste management, technology, and innovation.

While Hewlett-Packard provides the initial funding and is actively involved with each center for 3 years, it keeps long-term sustainability in mind, collaborating with local governments and agencies to ensure additional ongoing funding. It promotes the "train the trainer" model, involving local IT experts to work with members of the community to enhance digital literacy and also serve as role models.

Sources

"HP Digital Community Centres." Hewlett-Packard Development Company, Feb. 2006. Web. http://news.bbc.co.uk/2/shared/bsp/hi/pdfs/16_05_06_digital.pdf.

3.2.1.2 Cybercafés

Cyber cafes remain a very important public Internet access point around the world, particularly in emerging countries. A database tracking cyber cafes deployment compiles data on 4,208 facilities[5] broken down as follows (see Table 3.2):

While incomplete in terms of the total population, the statistics in Table 3.2 are important in terms of pointing out that even in countries with high broadband penetration, cybercafés remain quite popular. For example, the database includes 478 cybercafés in the United States and 149 in France alone.

---

[5] *Cybercafes* compiles only advertising locations, which means that it underestimates considerably the total population. Nevertheless, it provides a sense of the popularity of the access medium.

Table 3.2 Cybercafés by Continent and Region (2014)

Continent	Number	Percent (%)
Asia	1,091	25.9
Africa	175	4.2
Latin America and Caribbean	310	7.4
Europe	1,578	37.5
North America	859	20.4
Oceania	195	4.6
Total	4,208	

Source cybercafés

3.2.1.3 Local Area Network Houses

In developed and developing countries alike, privately owned local area network (LAN) houses not only promote broadband access, but also foster a community of online gamers who can connect and compete with each other. Many centers now offer digital training and other services as well, but in the least, they promote the social aspect of high-speed Internet use and increase the demand for such services. LAN houses typically consist of a network of connected computers where users can congregate to play the games, though many have expanded to offer additional services.

In some instances, they serve as the only point of access for many citizens, while in others they supplement household broadband access, serving a more social function. Most LAN houses charge users an hourly fee and the popularity of the houses keeps the prices down as owners compete with each other to attract more customers and drive business. They have been credited with driving digital inclusion, particularly important in countries with little public investment in broadband access and low penetration rates.

Case Study 3.34: LAN Houses (Brazil)

Throughout Brazil, citizens can access the Internet at Local Area Network (LAN) Houses, which took the country by storm in 1998. Each "house" consists of a network of assembled computers. Previously found exclusively in wealthy communities, the LAN houses are now most popular in poor, rural regions without easily accessible computers and broadband. While the houses were originally designed to support multi-player video gaming, many users report that they use the LAN Houses to stay informed and conduct job searches or work on school projects.

The LAN Houses reduce the country's digital divide, offering affordable ICT and broadband access regardless of location or socio-economic status. They have led to increased sociability and promote e-governance and e-education. Many also offer computer training courses and help with resume creation and job searches.

While owners choose their own pricing strategies, LAN houses typically charge between US$ 0.40 and US$ 1.50 per hour. Some neighborhoods have more than 100 LAN houses, many of which stand side by side.

The LAN houses came in part as a result of the federal government's Computers for All development project, which created credit lines allowing low-income families to purchase computers in small monthly installments. In some instances, citizens would purchase a computer and charge people to use it. As they accrued profits, they would purchase more computers and broadband access.

By 2008, the country held more than 90,000 LAN houses, accounting for half of all Internet access in Brazil and 79 % of all Internet access amongst the two poorest classes. For many of these citizens, the LAN houses were their only means of accessing the Internet. By 2010, an estimated 35 million citizens utilized LAN houses, a number slightly below previous years due to increased mobile phone penetration. This trend continued, as noted in a 2013 survey released by the Brazilian Internet Steering Committee. That said, LAN houses still offered access to 68 % of the population in the lowest income brackets. The committee concluded that the access points remain critical for digital inclusion.

Sources

Góes, Paula. "Brazil: Socio-digital Inclusion through the LAN House Revolution." *Global Voices*. N.p., 28 Sept. 2009. Web. http://global voicesonline.org/2009/09/28/brazil-socio-digital-inclusion-through-the-lan-house-revolution/.

Lemos, Ronaldo, and Paula Martini. "LAN Houses: A New Wave of Digital Inclusion in Brazil." *Information Technologies & International Development* 6 (2010): 31–35. Web. http://itidjournal.org/itid/article/view File/619/259.

"Brazil." *Freedom House*. N.p., 2013. Web. http://www.freedomhouse. org/report/freedom-net/2013/brazil#.UwgAOM0jtFA.

Case Study 3.35: PC Bangs (Korea)

The Korean model of PC Bangs inspired the Brazilian LAN Houses. These LAN gaming centers allow multiple users to play video games simultaneously for a low fee. Unlike the typical notion of a cyber café, PC Bangs do not merely offer Internet access, but also foster online and offline social gaming communities. When first established, the centers were described as being similar to a quiet library, but have since evolved into an "open, frenetic coin-operated setting." By 2008, Seoul alone housed more than 22,000 PC bangs. Even as household broadband penetration rates in the country increase, the popularity of the PC bang remains stable.

The PC Bangs are especially popular amongst the country's youth population, with many students spending their afternoons playing online games with their friends after school. The rates of play—typically between US\$ 0.40 and US\$ 2.00—are targeted for this age group, with many PC bangs offering rates that get incrementally lower with usage.

That said, because PC Bangs are typically open 24 h a day, the demographics change based on the time of the day. The average user in the mornings, for instance, is a male between the ages of 30 and 50. Many of these users also come to the room to day trade or search for jobs. In January 2014, the government issued a strictly enforced ban on smoking and some PC Bang developers began offering cleaner, trendier spaces, both of which could result in a shift in the average user demographics.

Korea saw its first Internet connections in 1994, and within a year, Internet cafes emerged to offer individuals the high-speed access previously afforded only by universities and research institutions. In 1997, the government ended the monopoly of the country's only telecommunications operator, Korea Telecom, which gave birth to competitors like Hanaro Telecom, which built its business on broadband connections. At the same time, software developer Blizzard Entertainment released its game StarCraft for Microsoft Windows, and many users felt that they could not play the game adequately without a broadband connection, thus increasing the demand for such services and giving rise to the PC Bangs. To drive demand further, many of the bangs also offered StarCraft competitions and events. Because the popularity of games drives the demand for the PC Bangs and vice versa, it is not uncommon to see partnerships between manufacturers and PC Bangs.

Sources

Huhh, Jung-Sok. "Culture and Business of PC Bangs in Korea." *Games and Culture* 3.1 (2008): 26–37. Sage Journals. Web. http://gac.sagepub.com/content/3/1/26.

"The PC Room Culture." *Play as Life*. N.p., 8 July 2010. Web. http://playaslife.com/2010/07/08/the-pc-room-pc-bang-culture/

Lee, Min-Jeong. "South Korean Videogame Players Decry Moves to Clean Up 'PC Bang' Lounges." *The Wall Street Journal*. Dow Jones & Company, 27 Jan. 2014. Web. http://online.wsj.com/news/articles/SB1000 1424052702303819704579320233203273934.

3.2.2 Economics of Shared Public Access Centers

In general, access to shared public access centers is free, with two exceptions. In the case of government or NGO-sponsored cases, SME access could be charged. In addition, if the center is a for-profit enterprise (e.g. cybercafé), user access is charged. An informal survey of cybercafés access rates around the world indicates that, while these for-profit centers are often presented as an affordable means of gaining Internet access, the rates remain significantly high for large portions of the population (see Table 3.3).

3.2.3 Ancillary Access Centers

Research conducted in some industrialized (e.g. United States) and emerging (e.g. Mexico) countries would indicate that broadband access centers located as adjacent to existing facilities (such as libraries, cultural centers, and youth centers) register higher usage and citizen engagement than standalone centers. The experience of this type of center will be reviewed, including the factors leading to a higher success rate, as measured by frequency of attendance and operational sustainability.

By offering broadband access and related training in youth and community centers as well as in libraries and even post offices, the initiatives have the following advantages:

- Citizens already utilize these venues and are comfortable coming here;
- Beyond the physical infrastructure, the centers have resources and staff, which means less overhead for the program;
- Many of the centers already offer broadband access; digital literacy promotion can be as simple as offering support or advice to users, but can also include formal workshops or one-on-one training;
- Depending on the staff at the center, training can incorporate basic skills (email, Internet searches) as well as skills more relevant to the workplace or center; and
- Programs can simultaneously address digital literacy, awareness, and access

In instances where government centers, like post offices, are used as points of access, they not only serve to provide the public with broadband access, but also promote e-Governance and allow the government to deploy online services and better connect with the community. Other instances work with centers as a result of partnerships between local, federal, and national governments and international corporations and grassroots organizations.

Ancillary access centers play a significant role in communities where the population has limited broadband access in the home. Similar to aforementioned initiatives, as citizens become more comfortable with the technology through the access afforded by these centers, they will likely pass down their learned skills to family members, friends, and coworkers.

Table 3.3 Cybercafé: cost per hour (in US$) (2012)

Country	Rate	Country	Rate
Algeria	1.00	Japan	7.50
Argentina	3.00	Kazakhstan	1.50
Australia	7.50	Kenya	2.28
Austria	6.00	Lebanon	2.75
Azerbaijan	1.10	Libya	2.25
Bangladesh	1.70	Mexico	2.25
Bhutan	4.20	Namibia	2.48
Bolivia	1.00	New Zealand	5.25
Brazil	3.45	Nicaragua	2.00
Canada	4.30	Nigeria	5.40
Chile	3.00	Pakistan	0.60
China	2.50	Panama	2.00
Colombia	3.00	Philippines	2.00
Cuba	1.50	Qatar	3.00
Egypt	1.50	Russia	3.00
Ghana	0.60	Saudi Arabia	6.60
Guatemala	1.50	Singapore	5.00
Haiti	2.50	Sweden	6.45
India	1.35	Tunisia	2.20
Indonesia	0.66	Turkey	0.50
Iran	3.00	United Kingdom	7.00
Israel	4.00	Vietnam	3.00

Source Daub (2012). Cost of Cyberliving, retrieved from www.ForeignPolicy.com/articles/2004/07/01

Case Study 3.36: Institute of Museum and Library Sciences (United States)

The United States' Institute of Museum and Library Services (IMLS) oversees grant making, policy development, and research to support the country's 123,000 libraries and 17,500 museums in providing public access to knowledge, culture, and learning. In June 2012, the IMLS awarded a $250,000 grant to WebJunction, a non-profit that offers online training and networking to library staff with a focus on workplace-applicable skills. The organization promotes library technology and management as well as increased public access in small and rural locations. With this grant, Web-Junction established a partnership with select state libraries, federal policy makers, and the national non-profit Connect2Compete, which provides free digital literacy training, discounted high-speed Internet, and low-cost computers to promote increased technology access and economic competitiveness.

This partnership facilitates digital literacy training at the libraries and will serve as a model for future initiatives in other states. The grant not only factors in the infrastructure costs, but also the expenditures related to planning and promoting the program. Ideally, with proper preparation, this project will spur future initiatives with support from the private and public sector.

Per IMLS, one-third of Americans do not have home broadband access. The foundation recognizes that low digital literacy rates act as a barrier. The grant promotes ICT skills at through public libraries, which have historically served as "the nation's de facto digital literacy corps." For millions of Americans, the only exposure to broadband is at the public library, so it follows that digital literacy training should occur here. Prior to this initiative, the United States' National Broadband Plan acknowledged the role that libraries could and should play in increasing digital proficiency in their respective communities.

Seventyeight percent of libraries already offer informal digital literacy training—helping patrons to use computers and perform basic Internet searches—and 38 % host formal workshops. This project will raise awareness and support of these endeavors and promote effective engagement and instruction.

Since its 2003 launch, WebJunction has reached more than 80,000 library staff members and has continued to make strides in advancing continuing education. In November 2013, it received a $250,000 grant from the Laura Bush twenty-first Century Librarian Foundation to create online learning content for public libraries.

Sources

"IMLS Announces Grant to Support Libraries' Roles in National Broadband Adoption Efforts." *Press Releases*. Institute of Museum and Library Sciences, 14 June 2012. Web. http://www.imls.gov/imls_announces_grant_to_support_libraries_roles_in_national_broadband_adoption_efforts.aspx.

"About Us." *WebJunction*. N.p., n.d. Web. http://www.webjunction.org/about-us.html.

"Our Mission." *Connect2Compete*. N.p., n.d. Web. http://www.connect2compete.org/about-us.

"New Grants to Support National Continuing Education for Libraries." *New Grants to Support National Continuing Education for Libraries*. N.p., 6 Nov. 2013. Web. http://www.webjunction.org/news/webjunction/IMLS-awards-grants-to-OCLC-to-support-national-continuing-education-programs.html.

3.2.4 Stand-Alone Public Access Centers

Complementing the review of ancillary access centers, the experience of stand-alone community access units will be reviewed, focusing on some of the "do's and don'ts's" that will enhance their success rate. In general, some of the practices for access centers comprise the following:

- Ensure that centers issue annual or semi-annual reports informing about activities being held, courses, results, topics taught, number of participants, etc.;
- Conduct internal evaluations of access centers every six months, measuring and comparing indicators such as number of visits per month, number of users per month, indicating gender, age, email accounts, blogs, and websites being created, etc.; and
- Make sure that qualified personnel design the training activities and train the users.

Some of these stand-alone mobile facilities take the form of an "Internet bus" which goes to the neighborhoods to provide training upon request. An effective measure of ownership is when a group neighbors, a club or an association make a reservation for the bus to come to their community.

Among best practices in the case of mobile access centers:

- Hold training sessions in small groups in a non-formal atmosphere;
- Faculty to student ratio should be 1:5, for sessions of no more than 10 attendees;
- Courses are typically held for 2 h and entail 5 sessions;
- Courses should follow a basic frame, that can be modified based on the knowledge level of participants; and
- It is advisable to manage mobile programs in coordination with a community library services which, in many cases, already operate library buses and stationary internet access centers.

Case Study 3.37: Community Technology Access Centers (Rwanda)

In 2009, the United Nations Refugee Agency (UNHCR) launched its Community Technology Access (CTA) project to give refugees and displaced individuals access to computers. The goal is to increase educational opportunities for this disadvantaged group, particularly women and girls. The access centers offer formal education, basic digital literacy course, long-distance learning programs, vocational training, and assistance with job applications. Further, access to Internet enables refugees to stay in contact with their family by way of email. In many ways, access to ICT tools will return some degree of stability to the refugee's lives.

Following training, refugees and members of the local community have the opportunity to work with the CTA as facility managers, technicians, and trainers. The program focuses on long-term sustainability, and hopes to create a design that will allow for seamless replication and scaling.

The project piloted in 2009 in Rwanda's Kiziba camp, where refugees fleeing the Democratic Republic of Congo had lived for more than 12 years at that time. Within the camp, refugees had very little access to educational opportunities and many had never seen—let alone used—a computer. With little likelihood of returning to their home countries, the refugees acquired the tools and skillsets necessary for self-sustainability, which came with increased access to and understanding of ICT.

The partnership between the UNHCR and corporate sponsors makes the program possible. As a proponent of renewable energy, UNHCR provides rural CTAs with solar power as necessary. Microsoft offers expert advice for the project and donates software while Hewlett Packard donates computers and hardware. PricewaterhouseCoopers offers pro bono staff and project management consulting and the Motorola Foundation provided the funding for the centers in Rwanda and Bangladesh.

UNHCR did not release information regarding the exact financial cost of the project, but its 2011 budget for all projects within Rwanda totaled US\$ 35.2 million.

Since the success of the initial 2009 pilot, the CTA program has expanded to include 31 centers in 13 countries: Azerbaijan, Armenia, Argentina, Bangladesh, Bulgaria, Georgia, Kenya, Mauritania, Nepal, Rwanda, Uganda, Sudan and Yemen.

Sources

"Computer Gateways to Self-Sufficiency." *Community Technology Access.* UNHCR, n.d. Web. http://www.unhcr.org/pages/4ad2e8286.html.

UNHCR-Microsoft Partnership Applying ICT to Support Refugee. Rep. UNHCR. Microsoft., June 2009. Web.

3.3 Advanced ICT Training

As a complement to the digital literacy programs examined above, this section will review initiatives regarding advanced ICT training, focusing on workforce development, development of capabilities of SME personnel, and creating awareness of the potential of broadband among government employees.

Beyond fomenting perfunctory skills, advanced ICT training can allow individuals to establish a career in the ICT industry, which typically offers higher quality and higher paying job opportunities. As such, this type of training ultimately allows a country to diversify away from an economy exclusively dependent on agriculture or basic manufacturing to one based on high value skills. Further, advanced ICT skills can permeate other industries, improving business practices in, for example, the healthcare and finance segments.

One of the most obvious targets for this type of training is colleges, universities, and certain vocational institutions. Depending on the level of training, programs can be offered as college majors geared toward employment in the IT sector or incorporated into a more general curriculum. Other programs may work with employees within a corporation or government workers. Given that many students or participants are likely older or already employed, advanced ICT training programs need to emphasize flexibility and convenience. While face-to-face training may work best in some instances, distance training and online programs may better suit other individuals.

Apart from programs integrated into the education curriculum, successful initiatives have included arrangements with government bodies, corporations, and international organizations. Emerging markets in particular have benefited from the leadership and direction of such organizations and partnerships with more advanced countries. While basic skills should remain a standard part of any ICT training program, more advanced courses may also include topics most relevant to the national economy.

Case Study 3.38: APCICT (Asia)

In June 2006, the United Nations Economic and Social Commission for Asia and the Pacific (ESCAP) inaugurated its United Nations Asian and Pacific Training Centre for Information and Communication Technology for Development (UN-APCICT/ESCAP). Located in Incheon, Republic of Korea, the Center aims to encourage member countries' efforts to increase ICT use. To do so, it promotes human and institutional capacity building by focusing on three key areas: training, research, and advisory services.

The region's developing countries suffer from lower ICT penetration rates than many of their developed counterparts. These rates, however, are not so much reflective of lack of access and connectivity or insufficient infrastructure, but rather of the fact that most countries do not have the technical human capital necessary to utilize these tools. The program, therefore, concentrates its efforts on advancing digital literacy to promote ICT use and understanding.

Launched in 2008, the APCICT Academy—APCICT's flagship training program—targets government leaders through its modular training program, promoting a higher understanding of the role ICT can play in development areas such as governance, education, health, business, and trade, amongst others. The Virtual Academy serves as its distance-learning platform and was established to maximize access to the academy's course materials and presentation. It also offers learning management tools and certifications. Users can access and download all Academy materials and course trainers can customize them to fit the needs of their training sessions. One module awards an e-certificate of completion to users who answer more than 80 % of final module quiz questions correctly. Print certificates are awarded to users who complete the final quizzes of all eight modules in the Academy of ICT Essentials for Government Leaders Program.

Beyond training programs, APCICT published its *Knowledge Sharing Series (KSS)*, which offers case studies and best practices research to raise awareness of and spread ideas related to ICT for development (ICTD). It also launched its online *Communities of Practice (CoP)*, a portal facilitating knowledge sharing amongst experts, field practitioners, and students.

To support its socio-economic development goals, APCICT engages with other UN bodies, government and non-government organizations, private sector corporations, and training institutions. These partnerships all work toward increasing national ICT capacity and providing government leaders with the ICT understanding and know-how necessary to reach their development objectives.

Since 2006, APCICT has rolled out capacity building programs in 29 countries throughout the region as well as additional programs in Africa with plans to extend to the Middle East. The Center's efforts have reached nearly 12,000 individuals through face-to-face and online trainings and it has worked with more than 80 national and regional partners from government, civil society, academia, and private sector institutions. APCICT has also published or contributed to over 90 ICTD resources.

Sources

"About Us." *UN-APCICT*. United Nations ESCAP, n.d. Web. http://www. unapcict.org/aboutus.

3.4 Small and Medium Enterprises

Building digital literacy among SMEs is directly linked to the improvement of a nation's economic performance. As reviewed by Katz (2012), broadband adoption contributes to economic growth. Considering that SMEs represent the majority of establishments, and the size of their participation in the national product in any given country, programs that target SMEs should have a significant economic return.

Broadband and ICT awareness development among SMEs is based on two distinct types of efforts: training of both personnel and management, and the provision of specialized consulting services to facilitate broadband adoption.

3.4.1 Training for SMEs

By its own definition, SMEs tend to be late adopters of broadband and ICT in general in the enterprise universe. In this context, the potential training initiatives focused on SMEs will be reviewed, ranging from efforts conducted within economic development units to industrial promotion administrations. Examples will be provided of types of programs and curricula that have proven to be particularly effective.

Because of its economic rationale, training on IT and broadband focused on SMEs is typically linked to large-scale national or regional initiatives. Some initiatives could entail formal certification of skills acquired, although others provide informal training on computer and Internet use. SME training programs are primarily focused on improving usability of the technology. As such, they tend to comprise standard computer courses and, in some cases, applications that fit onto existing enterprise work processes.

Given that these types of initiatives are economically focused, it might be pertinent to consider whether the programs should be offered free of charge or require some payment. In general, while training is provided for free, there are programs where users are asked to pay a symbolic fee, or even contribute to program expenses by acquiring training material (which in some cases could be reduced through a subsidy).

SME training programs should address the following concerns:

- Skill development: training should emphasize skills relevant to the workplace;
- Target audience: to the above point, the skills emphasized should focus on local relevancy (e.g. as they would relate to textile or coffee production in the case of a small village in Africa) rather than standard ICT skills;
- Convenience: offer training in areas that are easily accessible to the largest segment of the population; beyond physical workshops, consider the transmission of training through websites or video distribution to reach a larger audience that may not be able to attend the face-to-face classes due to time or location constraints; and
- Purpose: while these programs will ideally create a workforce capable of integrating ICT tools into the workplace, they must also effectively raise awareness of their benefits.

As SME training programs have similar objectives to basic digital literacy programs, it may make sense to link the two, with SME initiatives acting as extensions of more basic programs but tailored to the needs of business owners and employees. By enhancing the ICT awareness and capability of SMEs, successful initiatives can enhance competiveness and productivity while allowing SMEs to compete with larger MNCs that have the staff and resources to utilize ICT and broadband.

Case Study 3.39: District Business Information Centers (Uganda)

As a technical cooperation agency, UNIDO's activities fall into three categories: poverty reduction through productive activities, trade capacity building, and environment and energy. By introducing productive activities to countries, the organization in turn allows their citizens to find a path out of poverty. Customized services to this end include—amongst others—industrial policy advice, technology diffusion, and SME development.

By 2008, many African countries had not yet reached their Millennium Development Goals (MDGs) despite the approaching 2015 deadline. The organization found that these countries relied on the success of SMEs to strengthen their economies and reduce the socio-economic divide. UNIDO argued that in order to enhance their productivity and competitiveness, SMEs must see increased ICT access.

In 2006, UNIDO and Microsoft signed a Memorandum of Understanding that led to the development of the District Business Information Center (DBIC) program in Uganda. For its part, Microsoft developed ICT-related services, training, and awareness for the rural business community through its Digital Literacy program and SME training curriculum. The program focuses on providing the DBICs with the tools and skillset necessary for entrepreneurship development and sustainability. UNIDO trains two staff members per center—a business information officer and an ICT trainer—in technology and entrepreneurship, allowing them to access information on markets, customers, and technologies and open doors for business development. The project was financed by UNIDO and a grant from the Austrian Development Agency.

In 2007, the DBIC project received the Africa Investor Award in the "Best Initiative in Support of Small and Medium Enterprise Development" category. UNIDO boasts that, since implementation, it has offered ICT training and support; increased access to new markets, technologies and services; and increased competitiveness and productivity. By 2011, the initiative had trained 315 men and 226 women across industries such as food processing, textile, and coffee.

Sources

Unido.org. N.p., n.d. Web. http://www.unido.org/.

Promoting Public-Private Partnerships. Rep. UNIDO, Microsoft, Dec. 2008. Web. http://www.unido.org/fileadmin/user_media/Services/PSD/ICT/ GSA%20White%20paper%205.12.08.pdf.

"UNIDO: District Business Information Centres in Uganda." Microsoft, n.d. Web. http://www.microsoft.com/publicsector/ww/international-organi zations/projects/Pages/district-business-information-centres.aspx.

Real Impact for Better Development. Rep. UNIDO, Microsoft, 2011. Web.

Case Study 3.40: *Ecaustria* **and TELEFIT (Austria)**

Established by the Federal Ministry for Economy and Labor and the Austrian Chamber of Commerce, the *Ecaustria.at* website serves as an online information and communication platform. It falls under the Federal Ministry's "Let's e-biz" initiative and offers best practice case studies, information on starting a new e-business, and an annual awards section recognizing the best e-business and multimedia products in the country. It also provides e-learning and IT-training courses.

The government created the website to transform Austria into an Information Age leader by raising awareness and providing training. These resources demonstrate how an Internet connection can assist businesses in establishing relationships with customers and suppliers and in entering the global market. At the time of the project's inception, many large companies already conducted business over the Internet, with small and medium-sized enterprises acting as their suppliers. By learning how to use this tool, SMEs could more easily connect and become more competitive. website served to convince SMEs of the benefits of e-businesses.

To develop the *Ecaustria* platform, planners spent six months working with more than 300 experts from business, science, and administration industries, resulting in specialized services covering all topics relevant to e-business, such as e-content, e-employment, e-location, etc. The success of the project has been attributed to its sophisticated marketing techniques, which include traditional advertising, Internet and event marketing, and extensive public relations targeting the business-to-business segment. In 2002, the European Commission nominated *Ecaustria* for its exemplary project award.

Complementing the "Let's e-biz" platform, TELEFIT also serves as an information platform. The initiative targets SMEs in remote regions, offering videotaped trainings of its live "e-business road show," which takes place in urban areas and highlights best practices and guidelines to encourage SMEs to take advantage of the e-economy. In the first four years alone, more than 300,000 participants attended either in person or virtually. TELEFIT aims to prepare SMEs for the changes in telecommunications by providing the necessary tools and knowledge to profit from this new technology. The program addresses the specific needs of SMEs and also offers financial support as necessary.

TELEFIT's services are delivered through the Institute of Economic Promotion's network. The European Funds for Regional Development (EFRE) works with the Federal Ministry of Economic Affairs and the Economic Chamber of Austria to fund the project. Each event costs approximately €11,000. When TELEFIT first began, approximately 30 % of SMEs in the country operated online. Within 3 years, this number had increased to 80 %.

By 2014, TELEFIT had hosted 12 roadshows, attended by 50,000 participants.
Sources
Benchmarking National and Regional E-business Policies for SMEs. Rep. European Commission, 12 June 2002.
OECD SME and Entrepreneurship Outlook 2005. Paris: OECD, 2005. Print.
"Telefit Roadshow." *Enterprise and Industry.* European Commission, 4 Feb. 2014. Web. http://ec.europa.eu/enterprise/dem/initiatives/721/telefit-roadshow.

Case Study 3.41: SVEA (Sweden)

Established in 2000 by the Swedish Alliance for Electronic Commerce (GEA), the SVEA program aims to raise awareness of the benefits of e-commerce for SMEs. Targeting those businesses with little IT knowledge and no prior e-business experience, SVEA focuses more on the evolving business processes than the actual technology. By demonstrating the commercial advantages of engaging in e-business, the project encourages these companies to join the digital economy.

To achieve this goal, SVEA utilizes best practice research to demonstrate specific uses of e-commerce for companies in all sectors and regions. Within each SME, SVEA works with select "ambassadors" who can then train and educate other employees in their respective companies. By utilizing the train-the-trainer model, the benefits of the trainings are more easily diffused across organizations to reach a larger number of employees, who can then benefit from expertise in e-business application. Further, any member of the SME can access the SVEA database, which serves as a knowledge exchange tool and also offers analysis of the company's day-to-day operations and case histories. The program also utilizes conferences and seminars and works with IT solutions providers to demonstrate how e-business can positively impact SME operations.

Beyond these mechanisms, SVEA offers interactive, informal training over existing URLs that cover such topics as billing, exports, and logistics. Business associations, local consultants, partner companies, and research firms work with the program to offer these services and such partnerships allow SVEA to adapt more readily to the various demands of different SMEs. Funding for the project comes from private sector partners that have an interest in supporting SMEs. This partnership also serves to enhance the reputation of SVEA. The initial two-year budget for the SVEA project was €1.8 mn total, with an additional €400,000 set aside for the third year.

When the program first began, it aimed to educate 100,000 SMEs on the benefits of e-business and to see 30,000 utilizing this tool. In the first two years, the e-business increased from 26 to 54 %. The number of SMEs utilizing the Internet to purchase supplies reached 41 %, and nearly one-quarter of SMEs sold their products and services online.
Sources
Benchmarking National and Regional E-business Policies for SMEs. Rep. European Commission, 12 June 2002.

3.4.2 Consulting Services for SMEs

Complementing SME training, the provision of subsidized consulting services to SMEs has proven to be a powerful tool to increase awareness and promote adoption of broadband services and ICT, in general. Consulting can be provided by different entities: government-sponsored units, university students supervised by faculty (for practical training purposes), and pro-bono private sector firms. As opposed to the aforementioned training programs, however, consulting services tend to go beyond overarching basic training and awareness raising, focusing more on the implementation of ICT into individual SME business. In many instances, this implementation not only affords higher levels of productivity and competency, but also allows SMEs to enter the e-commerce arena.

These types of programs are especially popular in the European Union, which, by 2000, housed nearly 19 million SMEs. Examination of such programs has yielded the below recommendations:

- Employ independent consultants who can offer their unbiased opinions and bring their professional experience to the table;
- Programs that deploy consultants to individual SMEs should work to establish an overarching standardization to ensure quality and equity of service;
- Establish local centers in convenient locations where SME owners can seek tailored advice;
- Incorporate some degree of flexibility into the budget; as SME consultancy programs grow and more SMEs seek their services, expenditures tend to increase;
- To finance the project, partnerships with private entities (e.g. financial institutions or consulting firms) may take some strain off government resources; and
- Programs should also offer to SMEs such resources as databases of best practices information, online support staff, and a means for SMEs to communicate with and learn from each other.

An important aspect of consultancy programs is the ability to provide customized, tailored support to SMEs. Such services may include the establishment of a long-term business model based on the integration of ICT while others should target the needs of each SME to integrate such technology as necessary. These services should consider such factors as an SME's available resources, limitations, and future goals.

Case Study 3.42: eASKEL (Finland)

The Finnish Ministry of Trade and Industry developed the eASKEL program to enhance SME e-business development by increasing management competency, implementing e-business opportunities, identifying e-business needs, and producing a development program for companies. The service trains private-sector consultants to work with SMEs with little IT or e-business experience. The consultants spend 2–5 days with their clients developing an e-analysis and offering technology integration advice.

Consultants design strategic plans based on the individual needs and resources of each SME. Employing independent consultants allows the SMEs to benefit from professionals with extensive IT and e-business knowledge, who can also use their prior experience with other organizations in formulating their analyses. As a "Branded Expert Service," the program's standardized format ensures accountability and quality for all participating SMEs.

Prior to eASKEL, Finland's Ministry of Trade and Industry partnered with the Ministry of Agriculture and Forestry and the Ministry of Labor to create regional Employment and Economic Development Centers (T&E Centers). These fifteen centers located throughout the country offer businesses and individuals advice and development services. After the T&E centers reported that SMEs needed more support in utilizing e-business tools, the eASKEL program was piloted and designed to run from 2001 through 2007.

The project's budget depends on the number of participating companies, with €320,000 allocated for every 500 companies, the annual target. Each company pays approximately 35 % of the consultants' fees and the remainder is funded by the government's T&E Centers, which have a budget specifically for consultation services. The government ultimately subsidizes 85 % of the consulting fees associated with the project.

Sources

Benchmarking National and Regional E-business Policies for SMEs. Rep. European Commission, 12 June 2002.

Case Study 3.43: Opportunity Wales (United Kingdom)

With a focus on access, usability, quality of service, and cost, the Opportunity Wales program promoted SME support networks by providing e-commerce consulting and implementation. The Wales Trades Union Congress partnered with British Telecom and HSBC Bank in 1999 to address the attitudes toward e-commerce in Wales while raising awareness of the tangible benefits it can offer businesses. The project offered SMEs customized support based on each enterprise's individual needs; an advisor worked with the company to develop the necessary steps to introduce e-commerce into its business.

The initiative emphasized the importance of offering businesses one-to-one support by advisors with proper training and experience in implementing e-commerce models. Advisors' consultation services utilized the research and lessons learned from the replication and feedback of similar projects in other parts of the region. Prior to the commencement of the initiative, advisors and planners conducted in-depth consultation to ensure the incorporation of these lessons.

The e-commerce Innovation Center provided formal standardized training and accreditation to the advisors and also made available relevant resources and knowledge. A contact center and 24-h Internet portal provide a guide to best practices and Better Business Wales serves to offer advice as needed.

Opportunity Wales recognized that as SMEs become more comfortable with e-commerce, they will experience increased sales and higher levels of efficiency. In turn, the region will benefit from prosperity and increased employment opportunities. Research conducted in 2000 demonstrated that while businesses in Wales knew about e-commerce, few had actually utilized it. In-depth analysis showed little knowledge when it came to the implementation process, and the low Internet penetration rates in the region further hindered e-commerce uptake.

The National Assembly for Wales provided the initial funding for the project, though the partnership grew to include additional private sector, government, and university partners. The project first applied for funding from the European Union in 2000 and received approval from the Wales European Funding Office (WEFO) in 2001. WEFO supplied €17 mn of the €33 mn budget, with the remainder coming from partner organizations and the National Assembly. The majority of the budget went toward face-to-face SME consultation services and implementation support, with a very marginal amount left over for grant aid.

Opportunity Wales aimed to train 130 advisors and directly reach 35,000 companies within the first 3 years. Per its website, the campaign is no longer running, but by 2011, Opportunity Wales helped nearly 12,000 businesses.
Sources

Benchmarking National and Regional E-business Policies for SMEs. Rep.
European Commission, 12 June 2002.
"Opportunity for SMEs in Wales." *Opportunity Wales.* N.p., 23 Mar.
2011. Web. http://www.opportunitywales.co.uk/about/.

3.4.3 Broadband and New Firm Formation

Broadband technology is a contributor to economic growth at several levels. First, the deployment of broadband technology across business enterprises improves productivity by facilitating the adoption of more efficient business processes (e.g., marketing, inventory optimization, and streamlining of supply chains). Second, extensive deployment of broadband accelerates innovation by introducing new consumer applications and services (e.g., new forms of commerce and financial intermediation). Third, broadband leads to a more efficient functional deployment of enterprises by maximizing their reach to labor pools, access to raw materials, and consumers, (e.g., outsourcing of services, virtual call centers.). The study of the impact of broadband on economic growth covers numerous aspects, ranging from its aggregate impact on GDP growth, to the differential impact of broadband by industrial sector, the increase of exports, and changes in intermediate demand and import substitution.

In the area of productivity enhancements, it is logical to assume that productivity of information workers, defined as the portion of the economically active population whose working function is to process information (administrative employees, managers, teachers, journalists) depends directly on the investment in ICT capital (and particularly broadband). The studies conducted by this author[6] have, in fact, concluded that the larger the per cent of the workforce dedicated to information generation and processing is, the higher the proportion of capital stocks invested in the acquisition of ICT infrastructure (see Fig. 3.3).

Figure 3.3 and the corresponding regression coefficient indicate the existence of a direct relationship existing between the amount of information workers and IT capital investment in a given economy: as expected, the larger the proportion of information workers in a given the economy, the more capital is invested in information technology.

How can one theoretically explain the relationship between ICT and productivity? In his economics dissertation at Harvard University (1982), Charles Jonscher raised the hypothesis that if we can measure the micro-economic impact of ICT on firm productivity, then we should also be able to link the growth in informational occupations and the adoption of technology to improve their productivity at the macroeconomic level. According to Jonscher, economic growth logically leads to increasing complex production processes. In turn, complexity in

---

[6] See Katz (2009).

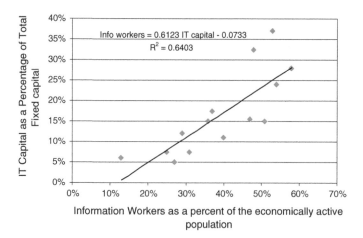

Fig. 3.3 Relationship between Information Workforce and IT Capital. *Note* Data for information workforce was derived from ILO statistics while IT capital was sourced from Nathan Associates (2011). *Source Adapted from* Katz (2010)

production processes results in increasing the functional complexity within firms (e.g. more inputs to be combined, more steps to be scheduled in a timely manner, more interactions occurring with suppliers of raw materials and with buyers of the end product). The first response of economic organizations to this effect is the creation of "information workers"—workers whose primary function is the manipulation of information for purposes of organizing the production of goods. At some point, however, information-processing workers become a bottleneck in the economic system. They cannot grow forever because this process reduces the overall availability of resources in other occupations. Furthermore, when information workers become a large proportion of the workforce, the complexity of information processing becomes a bottleneck itself. In other words, there is a limit to the possibility of manually storing, transferring and processing the growing amounts of information. This is where information and communication technologies come in. Their development and adoption is aimed at increasing the productivity of information workers and addressing this bottleneck. The availability of computing and communications allows firms (and their information workers) to be more productive in their manipulation of information. Broadband is a specific component performing this important productivity enhancement.

For example, research on the impact of broadband on productivity has successfully identified positive effects. For example, Waverman et al. (2009) determined the economic effect of broadband on the GDP of 15 OECD nations for the time period of 1980–2007. These included 14 European countries and the United States. By relying on an augmented production function derived from Waverman et al. (2005), the authors specified two models: a production function and a hedonic function for ICT capital stocks. Broadband impact on the productivity of the more developed nations in the sample was found to be 0.0013 and was statistically significant at the 5 % level.[7] In other words, Waverman estimated that for every 1 per cent increase in broadband penetration in high and medium impact income countries, productivity grows by 0.13 %. In another document, the authors commented upon the productivity effect in the countries of their sample with relatively low ICT penetration (Greece, Italy, Portugal, Spain and Belgium.). They found that broadband impact on productivity was nil, which indicated the high adoption costs, and critical mass thresholds.[8] In other words, for broadband to have an impact on productivity, the ICT eco-system has to be sufficiently developed.[9] It would appear, therefore, that in developed countries with high broadband penetration, the technology has an impact on aggregate productivity levels.

Beyond the aggregate impact on productivity, broadband has been found to have a contribution to accelerating the an economy rate of innovation, by promoting entrepreneurship and facilitating the formation of new businesses. These two areas of impact will be reviewed in turn.

3.4.3.1   Evidence that Broadband Facilitates Entrepreneurship

Beyond the employment and output impact of network construction, researchers have also studied the impact of network externalities on employment, derived from firm formation. This has been variously categorized as "innovation", or "network effects".[10] The study of network externalities resulting from broadband penetration has led to the identification of numerous innovation effects:

- New and innovative applications and services, such as telemedicine, Internet search, e-commerce, online education and social networking.[11]
- New forms of commerce and financial intermediation.[12]
- Mass customization of products.[13]

---

[7]   The original regression yielded a coefficient of 0.0027 for the 2/3 more developed countries in the sample and negative effect for the lower third. A negative effect did not make sense so the authors constrained the effect for the lower third to zero. At that point the coefficient for the full sample moved to 0.0013.

[8]   See Waverman 2009.

[9]   For example, Waverman et al. estimated that in the United States broadband penetration contributed approximately to 0.26 % per annum to productivity growth, resulting in 11 additional cents per hour worked (or USD 29 billion per year).

[10]   See Atkinson et al. 2009.

[11]   Op. cit.

[12]   Op. cit.

[13]   Op. cit.

- Reduction of excess inventories and optimization of supply chains.[14]
- Business revenue growth.[15]
- Growth in service industries[16]

However, for this to materialize, resources coming from business incubation programs set up by local governments, development agencies or multinational corporations are required. For example, mLab business incubators provide the following services[17]:

- Physical workspace
- Business training
- Tech training
- Testing and certification
- Business mentoring
- Market intelligence
- Competitions
- Funding
- Professional services
- Tech outsourcing
- Code repository

3.4.3.2 Virtual Business Incubation

The current conditions of digital innovation, particularly in emerging markets, reflect four shortfalls:

- Low liquidity (e.g. low depth of the market) flowing to innovation driven companies
- Low volume of innovation companies
- Lack of interest from domestic and foreign investors due to the market risk
- Low turnover of companies in terms of launching and decantation of the number of start-ups

At the highest level, business innovation in emerging markets faces supply side and demand side inefficiencies.

Supply side inefficiencies comprise limited availability of inputs. For example, start-ups are confronted with limited access to capital, due to the low availability of venture capital, and angel investors. As a result, small digital projects are difficult and costly to finance individually. In addition, start-ups might experience

---

[14] Op. cit.

[15] See Varian et al. 2002, Gillett et al. 2006.

[16] See Crandall et al. (2007).

[17] *MLab Business Plan.* Rep. InfoDev, Mar. 2011. Web. 12 Mar. 2013. http://www.infodev.org/en/Publication.1087.html.

limited access to technology infrastructure, and human capital. These supply-side innovation "failures" are of three types:

- Externalities in the innovation process: this means that it is common for a potential investor not to know the potential opportunities that might exist among innovative digital start-ups
- Information asymmetries: These work two ways

  - For liquidity, in a replication of the adverse selection and moral hazard problems highlighted in financial services, the lack of full information makes it hard for outside investors to evaluate the quality of the new ventures
  - For inputs, even when the infrastructure or inputs exist, innovative start-ups do not have information on their existence, which prevents them from making the right decisions to access the required development tools or infrastructure

- Coordination failures: limited access to inputs of other firms that would yield productivity enhancements; most
- Importantly, start-ups lack information on how to commercialize products/solutions/applications developed from innovation, how to ensure delivery (distribution channels) and actual utilization by citizens

Demand side inefficiencies revolve around the limited size of the local market. For example, the local market is not large enough to provide the necessary scale to affront the development of a technology firm. In this context, demand side inefficiencies can be overcome by export-oriented policies targeting outside markets. India and Malaysia address the technology demand gap by focusing on international software, IT outsourcing and technology markets Ireland and Israel also solve the size of the market inefficiency in software products by focusing on international markets (Ireland addressing MNCs, Israel targeting primarily the US).

Building a vibrant innovation-prone environment requires addressing these two types of gaps (see Fig. 3.4).

Broadband can have beneficial impact in the process known as virtual business incubation. As defined by Stam et al. (2011), virtual business incubators provide services beyond the confines of a physical building. This allows a start-up to use the services of an incubator, without actually being located at the incubator site, for instance through extension workers, online tools and off-site advisory services. They can also serve a much larger number of companies over an extended geographical area.

Virtual business incubation services, comprising business development support (training, mentoring), access to business networks and funding sources can be delivered through a variety of services:

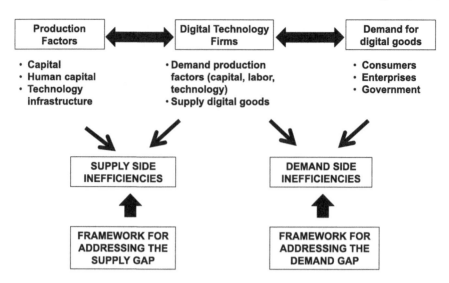

Fig. 3.4 Addressing innovation supply-side and demand side inefficiencies. *Source Developed by the author*

- On-site virtual services: group training, workshops, presentations and events at which start-ups can participate;
- Websites and E-learning: provide online training materials and information
- One-on-One exchange: traditional (E-mail, SMS, phone, Skype, and online collaboration tools);
- On-line staff recruiting: recruitment of mentors, staff, interns, investors
- Virtual communities and events: communities and online platforms where entrepreneurs meet, exchange knowledge and information and build collaboration networks; and
- Crowdsourcing and crowd-funding: support start-ups by mobilizing large numbers of knowledge and funding sources.

Broadband is a critical enabling platform that will facilitate delivery of these services. In fact, Stam et al. (2011) study of virtual incubators point out that unlimited supply of fixed and mobile broadband represents a critical component of these services. For example, a number of virtual incubators (Sofstart, BTI) have developed broadband connected satellite offices that can offer remote training or assistance to events to start-ups that are located in other geographies. Similarly, The *Business Links* website of the UK government offers a wealth of online information, including checklists, training materials and online instruction videos on practical issues, like business registration, hiring a new employee, or assessing and improving business cash flow.

3.4.3.3 Online Mentoring for Startups

A particular example of virtual incubation is the provision of online mentoring. In addition to access to finance and production inputs, start-ups are in great need of mentoring that helps them address the development and scaling up challenges.

Typical scaling up challenges encountered by start-ups include the aggressive growth of the customer base, assembling capabilities to support growth, modifying communications and decision making processes, functionalizing roles, and building cross-functional teams and processes. Given the limited experience often prevalent in start-up firms, mentoring becomes a critical requirement. Sources of mentoring are relatively scarce: experienced serial innovators, financiers, technology, and business executives. Mentoring includes counseling support in areas such as recruiting, organizational design, and professional development.

In this context, broadband can allow start-ups, particularly those based in remote locations, to gain access to the much-needed mentoring. The mentoring provided through on-on-one interaction tools saves time and travel costs and speeds up the provision of counseling, which can be critical in some contexts.

For example, the Founder Institute—a global network of start-ups and mentors—offers a four-month training program. Mentoring is provided through a global network of over 400 mentors, who are normally the CEO or founder of a successful start-up. The mentors are accessible both during the sessions and via email. Graduates can present their companies at so-called Founder Showcase Events in order to attract investors. Similarly, ParqueTec in Costa Rica has deployed a mentoring approach for remote (rural) strat-ups. This is based on one-on-one, regular face-to-face interactions combined with phone, SMS, E-mail and Skype contact.

3.5 Sponsorship Structures of Awareness Programs

Having reviewed the different awareness programs, this section will present the different sponsorship options regarding implementation and funding of such initiatives. The review of each option will include an examination of advantages and disadvantages, as well as observed best practices.

Digital literacy program sponsorship is a critical component of its success. Research conducted by Hilding-Hamann et al. (2009) found a direct relation between the sustainability of a digital literacy program and the sponsors involved. Sponsorship, as a term, entails more dimensions than just funding. It addresses the parties to be involved as stakeholders in the program. In general terms, they may include public organizations, educational institutions, private industry (both domestic firms and multinationals), non-governmental organizations, and local interest associations.

The sponsorship structure defines the contributions to be made by each stakeholder, and, in particular, the financing structure of the program. Key decisions to be made in this domain comprise the relative contribution of public and private funds, the in-kind (e.g. equipment) contributions, whether users should pay for participating in the program, and volunteer contributions. As mentioned above, the larger the group of stakeholders involved, the more likely is the project to remain sustainable. Partly, this results from financial support provided by private parties, which ensures long term funding being less subject to the vagaries of government budgeting processes.

3.5.1 Public Programs

Digital awareness and literacy efforts can be sponsored and funded directly by governments and/or other public entities. As found by Migliorisi and Balletti et al. (2009), broadband access can lead to increased productivity and development of human capital, but only when coupled with six basic "building blocks," such as basic computer skills. In fact, ICT skills—which range from the basic skills necessary for everyday life and work to industry-specific know-how—bolster the larger concept of digital literacy (Bowles 2013). UNESCO echoes this finding, viewing digital literacy as a "gate skill" that enables other life skills and that employers require (Karpati 2011).

Per the Australian government's Innovation and Business Skills (IBSA) council, the government can play a "critical role in both national advocacy and … [the formation of] a development plan and skills strategy to address digital literacy" (Bowles 2013). Because low levels of digital literacy tend to occur in disadvantaged regions, funding for programs that increase exposure to ICT training amongst rural and low-income members of society is critical in addressing the digital divide.

As the Broadband Commission (see Sect. 3.5.2.) asserts, governments need to make digital skills a priority in the education system to empower teachers and students and improve the overall quality of learning (The Broadband Commission 2013). Students armed with digital literacy skills can positively contribute to the workforce, in turn leading to increased economic activity. Further, digital literacy allows governments to extend their reach to and communication with their constituents through online government services.

Similarly, amongst the seven pillars of the European Commission's Digital Agenda is Pillar VI, which focuses on enhancing digital literacy, skills, and inclusion. Recognizing that 30 % of Europeans—mainly the elderly, unemployed, or low-income individuals—have never used the Internet, the European Commission calls for Member States to prioritize digital literacy to reduce disparities amongst the countries (Digital Agenda for Europe, n.d.). In doing so, the commission has undertaken such tasks as proposing indicators of digital and media literacy to establish a framework for benchmarking. Similarly, it has worked with Member States to implement digital literacy policies and promote awareness. The

European Union supports the agenda through its ICT Policy Support Program while also making funding available for related projects.

Along similar lines, the United States passed the American Recovery and Reinvestment Act (ARRA) stimulus plan in 2009 largely to create new jobs while encouraging economic activity and long-term growth. As part of this mission, the ARRA offers funding for digital literacy programs throughout the country. In other instances, governments have used the proceeds of spectrum auctions to fund digital literacy programs. In 2001 and 2002, the Portuguese government, as an example, raised €460 million when auctioning 3G mobile licenses, which it used to fund its educational technology plan (Trucano 2012).

3.5.2 Multilateral and Public Donors

Despite government efforts, the digital divide persists between developed and developing countries, demonstrating a need for intervention or assistance to increase ICT skills and access in the world's less connected countries. In recognizing the benefits of broadband diffusion, NGOs and multilaterals have increasingly shown an interest in and willingness to invest in digital literacy and awareness programs.

To strengthen the international broadband agenda, the ITU and UNESCO partnered in 2010 to launch the Broadband Commission for Digital Development. International agencies, government leaders, and private sector industry leaders comprise the group. The commission offered six overarching recommendations to address both access to broadband and education, amongst which included the recommendation to "teach ICT skills and digital literacy to all educators and learners" (The Broadband Commission 2013).

The 2013 *Technology, Broadband, and Education* report again stresses that, particularly in developing countries, barriers to broadband remain, with an examination of these countries discovering that they suffer from low levels of broadband literacy amongst all segments of the population (The Broadband Commission 2013). In working to promote universal broadband access, the Commission's projects many times highlight the technology's power to address the Millennium Development Goals and broader issues impacting society and development, such as gender inequality and lack of quality education. For example, introducing broadband literacy to the classroom can simultaneously improve education and create a higher skilled workforce while also empowering girls with ICT literacy.

In their 2013 *Aid for Trade and Value Chains in Information and Communication Technology* report, the OECD and World Trade Organization (WTO) reiterate the digital divide between developed and developing countries, emphasizing the relationship between ICT skills and economic development (Lanz 2013). While broadband penetration amongst the latter group has certainly increased, overall lack of Internet access and digital literacy continue to contribute to the digital divide between developed OECD countries and developing countries.

These skills can transfer to other areas of social development—including health, education, and women's empowerment—while enabling these countries to transition into more ICT services-focused economies.

When surveying 80 suppliers and 44 lead companies, the OECD-WTO report found that suppliers viewed lack of labor force ICT skills as one of the largest challenges faced by developing country suppliers (second only to access to finance). As such, the WTO's Aid for Trade initiative—which tackles trade-related issues in developing countries—works to prioritize ICT services with help from donors and partners, the majority of whom are part of ICT-focused public-private partnerships. Interestingly, however, the proportion of ICT projects comprising total Aid for Trade in many of the least connected countries is lower than the average amount seen in partner countries, which speaks for the role Aid for Trade and similar programs could potentially play in supporting ICT development.

Beyond multilateral support, many international and local foundations have formed to promote the digital literacy agenda. In some instances, NGOs receive private sector and government support of their endeavors. In 2013, for instance, IBM Ireland partnered with non-profit organizations to fund and launch the Socialcomputing.i.e. website, which encourages computer users to utilize online services and offers links to tutorials (Burke 2013). In another example, Microsoft partnered with SANGONet—a South African NGO network that focuses on ICT services—to create its NGO support program, which donates software to public benefit organizations (Microsoft 2014).

Ideally, NGO, multilateral, and government funding programs should work together to maximize resources for optimal efficacy and efficiency. An online survey of digital literacy experts from the educational, policy, and community roles in Australia, for instance, suggested that digital literacy programs would have more success when led by local councils and school boards (Bowles 2013). Even countries and cultures have different needs for digital literacy (UNESCO 2011), so local leadership could minimize the potential counterproductive factor of large-scale initiatives. As an example of such a partnership, with UNDP funding, local mobile ICT labs traveled to economically disadvantaged areas of Malaysia to promote digital literacy (Samudhram 2010).

Case Study 3.44: mEducation Alliance (International)

Comprised of 18 international organizations, foundations, and development banks, the Mobiles for Education Alliance works with multilateral and bilateral donors to improve formal and informal learning while making it more accessible, particularly in developing countries. The alliance reduces the barriers to education by promoting the use of low-cost mobile technology. To this end, the Alliance projects identify useful mobile applications that can address literacy, educational content, accessibility, educator development, and workforce training.

To support and advocate for m-education practices, the Alliance aims to do the following, per its website:

- Improve research, for example in the areas of mobiles for reading and content mastery via mobile delivery
- Perform a catalytic function by leveraging the Alliance's brand to promote and advocate for coordinated support of the effective use of mobiles in the administration and delivery of education
- Spread knowledge and awareness of the broad spectrum of available, accessible and affordable mobile technology tools available which can and/or have a proven positive impact on education outcomes
- Serve a convening function for practitioners, funders and leaders of the public and private sectors, to promote cooperation and coordination of efforts and knowledge in order to spur the innovation, affordability, and accessibility of mobile technologies for improved learning outcomes
- Explore opportunities to jointly fund specific activities of shared interests (e.g. evaluations of pilot projects, additional events focused on particular areas, etc.).

The Alliance also sponsors and backs a number of organizations and events that work toward its goal of improving education through mobile technology. Such funding may go toward international workshops and symposia, research roundtables, monthly seminars, and working groups. The mEducation Alliance website feature information about these events and related resources.

Many developing countries suffer from a lack of resources that in turn impact the availability of quality education. As mobile devices—including phones, e-readers, tablets, and projectors—become more prevalent, they create an opportunity to support education initiatives and reach citizens who may otherwise not have access to quality learning material.

Sources

"MEducation Alliance." *MEducation Alliance*. N.p., n.d. Web. http://www. meducationalliance.org/.

Case Study 3.45: Asian Development Bank (Uzbekistan)

In 2005, the Asian Development Bank (ADB) approved a regional technical assistance grant (RETA) targeting ICT investment in basic education within Central Asian countries. As part of this grant, the bank approved a US\$ 600,000 budget for Uzbekistan to implement an ICT in education strategy to increase its global economic competitiveness.

The project addressed not only digital literacy, but also ICT's potential to improve the overall quality of the education system and reduce the country's digital divide. The plan raised awareness of the benefits of—and barriers to—bringing ICT into the classroom while creating partnerships between the public and private sectors and international organizations.

In 2002, the country established a national ICT strategy to increase the computer-to-student ratio from 1:110 to 1:20 and to bring Internet access to 63 % of all schools by 2010. The ADB grant followed the United Nations Development Program's (UNDP) finding that Uzbekistan still lacked sophisticated ICT development. The research demonstrated that the country did not have enough computers to host computer classes or integrate new technology into the curriculum. Only half of all schools in the country offered computer classes, and many did not have access to any computers.

In addition to the ADP's US\$ 600,000 grant, the government of Uzbekistan contributed US\$ 150,000 toward the initiative. The Ministry of Education managed the project, which took place from March 2005 through February 2006. Beginning in 2006, the Government of Uzbekistan's National Program for Basic Education Development began another push to deploy Internet in classrooms. ADB provided US\$ 30 million of the US\$ 43 mn budget.

The first round of the project brought computers and relevant training to teachers and students at 300 "cluster" schools, with an additional 560 schools receiving computers in 2010. These schools served as resource centers for surrounding schools within a 30 km radius. Teachers from nearby schools could utilize cluster schools' resources while benefitting from shared training and collaboration. By 2010, more than 540,000 students directly benefited from the program while 90,000 teachers and staff received specialized training in the introduction of ICT into the curriculum. 25 % of the schools made the technology available to the public.

This project focused on disadvantaged students, with an estimated 165,000 students from poor and rural areas in grades five through nine benefiting directly. 70 % of the cluster leader schools were located in these areas. The initiative also served as a test in tackling Internet connections in rural classrooms through the use of mobile and wireless technology.

Sources

"Planned Project to Integrate ICT into Basic Education in Uzbekistan." Asian Development Bank, 10 Mar. 2005. Web. http://www.adb.org/news/planned-project-integrate-ict-basic-education-uzbekistan.

"High Tech for Young Minds." Asian Development Bank, 7 June 2010. Web. http://www.adb.org/features/high-tech-young-minds.

3.5.3 Public-Private Associations

These efforts comprise partnerships between public and private sector parties aimed at promoting broadband adoption. In setting up these associations, it is very useful to attract private parties, whose contribution is related to the core business of the participant. For example, in the Digital Communities Program in Dublin, Ireland, Hewlett Packard provides computer hardware, Eircom (the telecommunications carrier) provisions broadband connectivity, Microsoft supplies software and training support, the Dublin Institute of technology supplies staff, premises and administrative facilities, the Dublin Inner City Partnership contributes funding for salaries, and the Dublin City Council provides the premises for all the centers.

Best practices for managing public-private partnerships include:

- Set up an overseeing structure, such as a Management Board that meets regularly to discuss and manage progress of the program. The Board should comprise a senior executive from each of the partner's organizations, plus a representative from the community; and
- Additionally, the community should have a coordinator from each center, all of whom meet regularly to discuss issues faced in running their centers.

3.5.4 Private Efforts

While awareness efforts promoted by public sector are critical, they do not ensure automatic success. Administrations change due to electoral cycles and what could have been important for one party is not for another. Furthermore, public sector support does not necessarily mean unlimited funding. Finally, sometimes central governments are far removed from special groups to be targeted and, therefore, lack proper understanding of their specific needs.

In that context, support from the private sector might prove beneficial to improve sustainability of demand stimulation programs. For example, ZeroDivide is a philanthropic organization that seeks to increase digital inclusion in low income, mostly non-white communities in the United States. The program comprises a number of projects, such as training community members in the use of technology, increase household computer ownership through the provision of free or low-cost equipment, and develop community-focused content. The projects also included deploying Wi-Fi broadband networks and a community technology center for training and Internet access.

The Digital Inclusion project is a similar program. In this case the private non-profit organization partners with community organizations to distribute low-income households an ultra-portable laptop, high-speed broadband access, couple with digital literacy training, and content aimed at low-income households.

There are two types of benefits potentially derived from private efforts in broadband awareness. First, local companies can provide not only funding but also a good understanding of the needs of local groups. Secondly, multinational corporations can provide funding but also the possibility of cross-fertilizing experiences from one country to another in terms of "what works and what doesn't".

Case Study 3.46: Pasha Centers (Kenya)

In response to the World Bank's Regional Communications Infrastructure Project (RCIP), the Government of Kenya developed a program in 2007 to create "digital villages" to connect rural areas to the Internet. The World Bank suggested that Kenya serve as a "proof-of-concept" country to determine the feasibility of digital villages in other regions. The Kenya ICT Board (KICTB) partnered with the Cisco Internet Business Solutions Group (IBSG) to outline the key challenges and the solutions needed. Cisco then established a toolkit to implement digital villages in other emerging markets, focusing on villagers' needs and wants, the required capability, and the logistics behind future service and management.

Despite private-sector interest in these villages, Cisco recommended that the KICTB should first run a pilot program to understand fully its implications. The board agreed that rolling out digital villages prior to testing the initial concepts could prove disastrous. Thus, "Pilot Pasha Centers" (PPCs)—the Swahili word meaning "to inform"—were launched in January 2009. The committee selected 5 cyber cafes in rural areas across the country to serve as a test bed for the research that would build the model for large-scale deployment. By April 2010, these 5 PPCs featured 512 K connectivity, Cisco WebEx online conferencing, surge protectors, and various ICT equipment. Cisco also provided content from its Connected Knowledge Centers program. Following an initial third-party evaluation, the KICTB identified the factors necessary in ensuring Pasha Center success.

- Physical infrastructure such as reliable power supply and connectivity
- Auxiliary services
- Marketing for awareness and education
- Entrepreneurial initiative in customer service and experience
- Innovative new uses of the Internet for business collaboration and "edutainment"
- Training accreditation for various vocational e-learning courses

Cisco then developed tools to establish a business-planning model, determining that Pasha Centers could break even by generating an income of US$ 550 per month by offering the use of five computers for 8 h a day and charging one cent per minute of use.

Following the PPC program, Cisco developed a toolkit for Pasha Center managers and used the knowledge gained to scale and roll out a larger digital villages program. Kenya's Ministry of ICT announced in March 2010 that four service providers must each roll out five digital villages per constituency, totaling 4,200 centers. In 2012, the KICTB set aside a US\$ 315,000 loan for 26 entrepreneurs looking to build pasha centers. In the first six months, Kenya saw the creation of 63 new pasha centers, which provide approximately 30 % of the country's total ICT coverage.

Sources

Drury, Peter. *Kenya's Pasha Centres: Development Ground for Digital Villages.* Rep. Cisco, Jan. 2011. Web. http://www.cisco.com/web/about/ ac79/docs/case/Kenya-Pasha-Centres_Engagement_Overview_IBSG.pdf.

Gichane, Charles. "Kenya: ICT Board Loans Sh27 Million for Pasha Centres." *AllAfrica.* N.p., 11 July 2012. Web. http://allafrica.com/stories/ 201207120031.html.

Case Study 3.47: Computer Literacy and Training Program (India)

In recent years, India's economy has transformed as a result of the rapid growth seen in its IT sector. Paradoxically, by 2004, many of its citizens still lacked access to this technology, in large part due to geographic barriers. The nation's rural villages, for instance, tended not to have basic telecommunications infrastructure. At this point, IBM partnered with Department of Information Technology (DIT) in West Bengal—an agriculture-dependent state in Eastern India with just under 100 million inhabitants—to implement the same IT workforce-training programs that the corporation had deployed in countries such as Venezuela, China, and Egypt.

At this point in time, the DIT already made commitments to increasing IT access and training within its education system at the middle school and high school levels to prepare students to enter the workforce with the skills necessary for IT-related careers. The initiative, known as the Computer Literacy and Training Program, was credited with the region's sector growth and highest IT revenues in the country, but economists feared that the availability of skilled labor would not match the demand for such services for much longer. By investing in the training of all students, the government hoped to narrow the digital divide and also create the workforce supply necessary to strengthen its high-tech outsourcing economy.

To enhance the program, IBM Learning Solutions brought IT infrastructure, support, and management as well as education services to 400 schools in the state, which then aimed to reach more than 150,000 students within the first 3 years. Within each school, select teachers received IT training and certification so that they could act as instructors for students and other faculty members. Each school received 10 computers equipped with Intel processor servers and Red Hat Linux 8.0.

Those students interested in more high-level careers had access to advanced IT training. All instruction took place face-to-face in the local language across multiple platforms. Students took part in annual assessments and received formal completion certificates. IBM also offered orientation sessions for teachers who could continue the program following the end of the corporation's contract with the Government of West Bengal. The program was sustained by charging students US$ 0.75 per month.

IBM mangers keep the project on track through mechanisms such as delivery milestones, user satisfaction, and performance parameters while ensuring an adherence to class schedules, machine uptimes, and annual exams. The managers also take responsibility for keeping the government informed through reports and briefs.

By 2009, the program trained nearly 6,200 teachers in 330 private and public schools across 19 districts and enrolled 160,000 students. Evaluations have attributed an increase not only in IT skills amongst students, but also in overall academic performance as well to the program.

Sources

"Government of West Bengal Conquers Digital Divide with Help from IBM." IBM Learning Solutions, Dec. 2004. Web. http://www-935.ibm.com/services/us/imc/pdf/cs-west-bengal.pdf.

Survey of ICTs for Education in India and South Asia, Case Studies. Rep. PricewaterhouseCoopers, 2010. Web. www.infodev.org/en/Document.873.pdf.

Case Study 3.48: Google 'Good to Know' Campaign (United States)

In early 2012, Google launched its "Good to Know" campaign focusing on educating consumers about web-related privacy issues and the ways they can make the experience safer and more secure. Topics include privacy and security tips, such as how to use two-step verification, the way to lock a public computer, and how to make sure website connections are secure. Google not only published a Good to Know book on its website, but also ran ads—all of which are accessible for download on the website—in newspapers and magazines, websites, and subway cars in the New York and Washington, D.C. metro areas. The website covers four main: "Stay safe online," "Your data on the web," "Your data on Google," and "Manage your data." It also features sections that focus on online safety for the family and offer resources such as explanations of technical jargon, links to related Google services, and a list of organizations dedicated to providing help and advice online.

While informative, the ads are written in an entertaining, light-hearted manner, but with enough condescension to make readers realize that they need to pay more attention to how they use the Internet. Google's director of privacy described the campaign's target audience—the casual Internet user who may not be as savvy when it comes to online safety as he or she should be—in a corporate blog post. "Does this person sound familiar?" she asked. "He can't be bothered to type a password into his phone every time he wants to play a game of Angry Birds. When he does need a password, maybe for his email or bank Website, he chooses one that's easy to remember like his sister's name-and he uses the same one for each Website he visits. For him, cookies come from the bakery, IP addresses are the locations of Intellectual Property and a correct Google search result is basically magic."

Prior to its launch in the United States, Google first debuted the multi-million campaign in October 2011 in the United Kingdom through its partnership with the Citizens Advice Bureau. The campaign has continued to launch in global locations.

Sources

Boulton, Clint. "Google 'Good to Know' Campaign Touts Web Privacy, Security." *EWeek*. N.p., 17 Jan. 2012. Web. http://www.eweek.com/c/a/Security/Google-Good-to-Know-Campaign-Touts-Web-Privacy-Security-706900/.

"The Good to Know Campaign." *Google*. N.p., n.d. Web. http://www.google.com/goodtoknow/campaign.

References

Atkinson, R., Castro, D., Ezell, S.: The digital road to recovery: a stimulus plan to create jobs, boost productivity and revitalize America (2009). http://archive.itif.org/index.php?id=212. Accessed 23 Mar 2014

Belo, R., Ferreira, P.: Spillover effects of broadband in schools and the critical role of children (2012). http://www.academia.edu/2850670/Spillover_Effects_of_Broadband_in_Schools_and_the_Critical_Role_of_Children. Accessed 22 Mar 2014

Bowles, M.: Digital literacy and e-skills: participation in the digital economy. IBSA (2013)

Burke, E.: NGOs, IBM and businesses team up to tackle digital literacy (2013). http://www.siliconrepublic.com/careers/item/32379-ngos-ibm-and-businesses-te. Accessed 26 Mar 2014

Cisco.: Courses and certifications—networking academy (n.d.). http://www.cisco.com/web/learning/netacad/course_catalog/index.html. Accessed 6 Mar 2013

Crandall, R., Lehr, W., Litan, R.: The effects of broadband deployment on output and employment: a cross-sectional analysis of U.S. data (2007). http://www.brookings.edu/research/papers/2007/06/labor-crandall. Accessed 23 Mar 2014

Digital Agenda for Europe.: Pillar VI: enhancing digital literacy, skills and inclusion (n.d.). http://ec.europa.eu/digital-agenda/en/our-goals/pillar-vi-enhancing-digital-literacy-skills-and-inclusion. Accessed 26 Mar 2014

Daub: I could only find it for 2004: http://www.foreignpolicy.com/articles/2004/07/01/cost_of_cyberliving (2012)

Gillett, S., Lehr, W., Osorio, C., Sirbu, M.: Measuring Broadband's Economic Impact (2006). United States Department of Commerce, Washington

Goldfarb, A.: The (teaching) role of universities in the diffusion of the Internet. Int. J. Ind. Organ. **24**(2), 203–225 (2006)

Goolsbee, A., Klenow, P.J.: Evidence on learning and network externalities in the diffusion of home computers. J. Law Econ. **45**(2), 317–343 (2002)

Hauge, J. A., Prieger, J. E.: Demand-side programs to stimulate adoption of broadband: what works? Rev. Netw. Econ. **9**(3), 81 (2010)

Hilding-Hamann, K., Nielsen, M., Pedersen, K.: Supporting digital literacy: public policies and stakeholders' initiatives (2009). The European Commission on Digital Inclusion, Copenhagen

Horrigan, J. 2014. *The essentials of connectivity*. [report] Virginia. Comcast Technology Research & Development Fund

Hudson, H.E.: Communication Satellites. Free Press, New York (1990)

Hudson, H.E.: From Rural Village to Global Village. Lawrence Erlbaum Associates, Mahwah (2006)

Jonscher, C.: Productivity Change and the Growth of Information Processing Requirements in the Economy: Theory and Empirical Analysis. Harvard University, Harvard (1982)

Karpati, A.: Digital Literacy in Education. Policy Brief. UNESCO, Moscow (2011)

Katz, R.L.: El papel de las TIC en el desarrollo. Propuesta de America Latina a los retos economicos actuales. Madrid: Ariel (2009)

Katz, R.L.: Data requirements to measure the economic impact of broadband. Paper presented at the 8th World Telecommunication/ICT Indicators Meeting of the International Telecommunication Union, Geneva, Switzerland, 24–26 Nov (2010)

Katz, R.L.: The Impact of Broadband on the Economy: Research to Date and Policy Issues. International Telecommunications Union, Geneva (2012)

Lanz, R.: Aid for trade and value chains in information and communication technology. OECD and WTO (2013)

Microsoft.: Microsoft citizenship NGO support programme (2014). http://www.microsoft.com/southafrica/citizenship/ngo-support.aspx. Accessed 26 Mar 2014

Microsoft.: Microsoft innovation center activities (n.d.). http://www.microsoft.com/mic/mic-activities.aspx. Accessed: 6 Mar 2013

Migliorisi, S., Balletti, A., Edwards, K., Dona, R., Mendoza, D., Dias, U. S., Collado Di Franco, M., Deza, L.: Economic development and inclusion through local broadband access networks (2009). Inter-American Development Bank, Washington

Nathan Associates: *The economic and societal benefits of iCT use: an assessment and Policy road- map for latin America and the Caribbean*. bArlington, VA (2011)

OFCOM.: Adults Media Use and Attitudes. OFCOM London (2012)

Puma, M. J., Chaplin, D.D., Pape, A.D.: E-rate and the digital divide: A preliminary analysis from the integrated studies of educational technology. The Urban Institute, Washington (2000)

Samudhram, A.: Building ICT literate human capital in the third world: A cost effective, environmentally friendly option. Curriculum, technology and transformation for an unknown future. Proceedings Ascilite Sydney, pp. 844–851 (2010)

Stam, N., Buschmann, S.: Lessons on virtual business incubation services (2011). http://www.visbdev.net/visbdev/fe/Docs/VBI.PDF. Accessed 23 Mar 2014

The Broadband Commission Working Group on Education.: Technology, broadband, and education: advancing the education for all agenda. UNESCO, Paris (2013)

Trucano, M.: Around the world with Portugal's eEscola project and magellan initiative (2012). http://blogs.worldbank.org/edutech/portugal. Accessed 26 Mar 2014

UNESCO.: Digital literacy in education (2011). http://unesdoc.unesco.org/images/0021/002144/214485e.pdf. Accessed 22 Mar 14

United States Department of Commerce.: Fact sheet: digital literacy (2011). http://www.commerce.gov/news/fact-sheets/2011/05/13/fact-sheet-digital-literacy. Accessed 22 Mar 2014

Varian, H., Litan, R.E., Elder, A., Shutter, J.: The net impact study. Cisco Systems Inc, San Jose (2002)

Waverman, L.: Economic Impact of Broadband: An Empirical Study. [report] LECG, London (2009)

Waverman, L., Meschi, M., Fuss, M.: The Impact of Telecoms on Economic Growth in Developing Countries. Vodafone Policy Paper Series. GSM, London (2005)

Witherspoon, J.P., Johnstone, S.M., Wasem, C.J.: Rural Telehealth: Telemedicine, Distance Education, and Informatics for Rural Health Care. WICHE Publication, Boulder (1993)

Chapter 4
Achieving Affordability

This chapter focuses on one of the three dominant adoption obstacles needed to be tackled to achieve broadband diffusion among residential subscribers: limited affordability. The research reviewed in Chap. 2 indicated that, while price does not play a significant role among early adopters of broadband, once service coverage has reached a tipping point, affordability becomes the most important variable driving penetration. Moreover, research also indicated that, at higher penetration levels of broadband, price elasticity coefficients start to decline, indicating the lower importance of affordability as an adoption factor. In that regard, research indicates that limited affordability is a critical adoption obstacle when broadband penetration ranges between 3 and 20 %, which is the stage at which most emerging countries are.

In Sect. 4.1, the economics of broadband adoption will be presented. The section will introduce all the components driving the total cost of ownership of the technology. They comprise device acquisition and other on-time costs, service subscription retail pricing (with multiple subcomponents), and service taxation. This introduction will serve as a basis to discuss the potential policy and private sector initiatives addressing the broadband affordability obstacle.

Three types of initiatives, targeting the affordability obstacle will be reviewed (see Fig. 4.1).

Section 4.2 will review the service pricing obstacle and policies to tackle it. Service pricing will be discussed in terms of fixed and mobile broadband independently. It will first introduce an approach for conducting comparative pricing analysis, followed by presenting models of service price elasticity, ending with a review of potential private sector and public policy initiatives to reduce service pricing.

Section 4.3 will turn to device pricing. Broadband access requires devices capable of accessing the Internet. They range from computers supplemented with a modem (called USB modem, dongle, or air card) to smartphones, netbooks, and tablets. Since pricing dynamics (and capabilities) vary greatly across devices, the demand structural factors linked to device access will be discussed in two distinct sections: personal computers and mobile devices. In this context, several potential programs aimed at reducing device pricing will be presented.

R. L. Katz and T. A. Berry, *Driving Demand for Broadband Networks and Services*, Signals and Communication Technology, DOI: 10.1007/978-3-319-07197-8_4, © Springer International Publishing Switzerland 2014

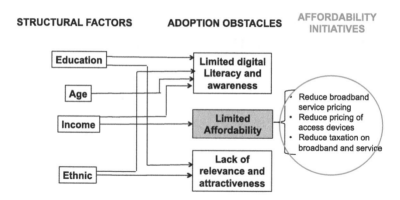

Fig. 4.1 Affordability policy initiatives in residential broadband

Section 4.4 will focus on taxation. This area will review the different equipment levies (import duties, value-added taxes, and sector specific) and service taxes (value added and sector specific). This review will serve as a basis to discuss the impact of taxation on total cost of ownership of broadband, and present examples of policy initiatives to tackle broadband taxation.

4.1 Economics of Broadband Adoption

The economic structural factors driving broadband purchasing need to be assessed in terms of the total cost of ownership, a concept that allows factoring in purchasing cost of devices, initial activation costs, as well as the recurring costs resulting from monthly service charges. Total cost of ownership is the sum of the cost of usage (service) plus part of the cost of the access device, which is assumed to be amortized throughout its lifetime, usually between 2 and 4 years, depending on the device.[1]

As a general principle, telecommunications services have negative elasticities: higher prices imply lower demand. However, pricing needs to be decomposed among its different elements because they affect broadband initial adoption and usage in different manners. Initial adoption is constrained by device acquisition, its corresponding tax burden, service activation cost, and expected recurring costs derived from subscription retail fees and taxes.

Device retail prices and their corresponding taxes vary between fixed and mobile broadband. Fixed broadband requires the acquisition of a personal computer (or a tablet with Wi-Fi access), while mobile broadband could be supported through either a personal computer, a smartphone, or a tablet. Retail acquisition

---

[1] The average replacement cycle of a personal computer is 3.5 years, while the cycle for a mobile handset is 18 months.

prices of this type of equipment are driven by supply and demand conditions, in particular manufacturing economies of scale and component costs. While device retail pricing is typically out of the realm of policy control, without an explicit industrial policy taxation is not. Final price of devices is affected by a set of different taxes, which vary by country and year. As will be shown below, taxes can, in some cases, add a significant burden to the retail price.

The importance of expected recurring costs on initial adoption varies by type of device. For example, in general, service subscription for a wireless modem is generally stable, representing a monthly rate for an expected type of service plan. As a result, subscribers can easily factor in the monthly subscription cost on the total cost of ownership and make an informed decision regarding adoption. In the case of smartphone access to mobile broadband service, prepaid or capped subscriptions allow subscribers to purchase service while limiting total cost of ownership to a minimal amount. In this case, taxes play a more limited role than in the case of service activation and postpaid usage rates.

4.2 Broadband Service Pricing as a Barrier to Adoption

In Sect. 2.2, the structural factors and obstacles affecting residential broadband penetration were reviewed. Among them, limited affordability was highlighted as a critical factor, according to which certain portions of the population either cannot acquire a device or purchase the subscription needed to access the Internet.

This section will focus on broadband service pricing, analyzing the impact of service activation and monthly subscription on demand. It begins by reviewing the current situation and trends regarding broadband pricing. At the same time, it presents a set of approaches for comparing broadband prices. Based on pricing data, it provides an analysis of price elasticities and presents tools for estimating increases in service penetration based on price reduction. The analysis of price elasticity will set the context to the review of different policy initiatives to reduce service pricing, as a way of stimulating adoption.

4.2.1 Cross-Country Comparisons of Fixed and Mobile Broadband Pricing

Broadband pricing does not lend to easy comparisons within and across countries. Operators tend to include different components in the price structure, ranging from speed (download and upload), limits on content download (known as CAPs), hardware costs (such as the modem and router), and activation costs (including installation charges). In addition, monthly prices can include voice subscription charges if broadband is offered within a service bundle. As a result, comparative analysis needs to normalize all of these components, in addition to control for purchasing parity differences across world regions. This section provides examples of pricing differences and presents an approach for developing price comparisons useful for serving as an input for formulating policy initiatives.

Cross-country comparisons of broadband prices require implementing a consistent approach for collecting pricing information and a methodology for normalizing/standardizing data. In the first, domain, the OECD has generated a methodology[2] for gathering broadband prices that comprises the following recommendations:

- If possible, data should be collected for three country operators (the incumbent telecommunications operator, the largest cable provider (if there is cable coverage) and one alternative provider, if available) offering service over DSL, cable or fiber. Offers should be advertised in the respective operator websites, be available in the country's largest city (or region if it is a regional carrier), and be communicated as monthly subscriptions.[3] It is usually the case that advertised speed is not similar to service throughput. Given the difficulty in collecting reliable service quality data, it is preferable to develop comparisons based on advertised speeds (although it would penalize the higher quality providers)
- In collecting pricing data, the following elements should be addressed:
  - Is broadband pricing a stand-alone offer or is it sold as part of a service bundle including voice communication, and/or television service?
  - Do offers include discounts for long-term commitments?[4] The OECD methodology recommends that, for standardization purposes, only commitments of 24 months or less should be included
  - Treating voice components of a bundle: Some plans offer a number of phone calls as part of the broadband plan, discounts for carrier preselection, or require a certain amount of phone use per month. The OECD methodology typically tries to exclude, if possible, all ancillary pricing elements from the broadband price
  - Finally, while the OECD methodology converts pricing data to purchasing power parity (PPP), it is worth mentioning that if comparisons are made within a certain region, depicting similar income per capita levels (e.g., Latin America), this conversion is not needed

---

[2] http://www.oecd.org/internet/broadbandandtelecom/criteriafortheoecdbroadbandpricecollections.htm

[3] While the OECD methodology excludes all offers under 256 kbps, a comparison across emerging countries might require including below 256 kbps products.

[4] In some countries, operators offer significant discounts to subscribers who commit for long periods. These longer term subscribers can often comprise a large majority of all broadband subscribers in a country (e.g., in the Republic of Korea it is possible to benefit from further discounts for commitments over 24 months).

- In collecting data for comparing CAPs (the limit on information to be down-loaded from the Internet), the following recommendations should be followed:

  – Present CAPs in megabytes per month
  – CAPs should be compiled for all domestic and international traffic. How-ever, in cases where national and international traffic are capped differently, the caps for international traffic should be considered for comparison
  – Costs for additional traffic should be accounted for in price per additional megabyte
  – When operators offer additional monthly traffic in different bundles the price should reflect the lowest price per MB across offers
  – When prices per additional megabyte are on a graduated scale, the average of all prices should be used
  – Prices and bit cap measures should not take into account bandwidth offered during specific times of day

Once data is collected, its normalization for comparison purposes needs to follow a set of standard procedures:

- Use average pricing for the first 24 months of service purchase
- Use a common set of offers:

  – For fixed broadband, start with a basic plan with a 2 GB cap
  – However, in some countries the basic plan (which is offered at a low download speed) could have been withdrawn from the market; in that case, it could be preferable to measure a service that is likely to stay in the market for an extended period of time: 2.5 Mbps download speed and 6 GB cap. This choice normalizes product selection eliminating the low performance products that could be offered in certain countries as the "Basic" plan
  – The OECD recommends (OECD 2012) the usage of broadband price bas-kets for comparison purposes. For example in fixed broadband, basket 3 is defined as services with at least 6 Mbps of download speed and a 6 GB cap. In the case of mobile broadband, the OECD differentiates between laptop, tablet, and handset use, proposing five baskets per device (see Table 4.1).

4.2.2 Broadband Price Elasticity

As reviewed in Sect. 2.2.4, broadband price elasticity is a function of service adoption. At lower levels of service adoption, broadband is price inelastic. This means that early adopters are not sensitive to price declines. Beyond a threshold point near 3 % adoption, price elasticity increases significantly and persists at high levels up to 20 % penetration, when it starts declining again.

Evidence of this behavior has been identified in a number of studies of fixed broadband pricing. On the other hand, research on mobile broadband pricing is not

Table 4.1 Mobile broadband basket proposal—tablet, laptop and handset use

Laptop use (data volumes)	Tablet use (data volumes)	Handset use (data volumes + voice/SMS basket)
500 MB	250 MB	100 MB + 30 calls basket
1 GB	500 MB	500 MB + 100 calls basket
2 GB	1 GB	1 GB + 300 calls basket
5 GB	2 GB	2 GB + 900 calls basket
10 GB	5 GB	2 GB + 100 calls basket

Source OECD Broadband Portal

conclusive as of yet in terms of estimating service price elasticity. This is due to the technology's recent deployment, compounded by the complexity of estimating cross-elasticity between fixed and mobile broadband (two services that will be increasingly becoming substitutes).

The following section reviews evidence of price elasticity in each service and presents a model that could serve as a tool for estimating service adoption as a result of potential price reductions.

4.2.2.1 Fixed Broadband Price Elasticity

There are several studies that shed a light on the potential price elasticity of broadband services. For example, by relying on data from a survey of approximately 100,000 US households, Goolsbee (2006) found fixed broadband service demand between 1998 and 1999 to be fairly elastic. According to the author, for levels of penetration between 2 and 3 %, price elasticity was between −2.8 and −3.5.

In another study, Galperin and Ruzzier (2013) utilized data for the OECD and Latin America to estimate the elasticity of fixed broadband service in 2011. Their main finding is that the elasticity of both regions varies. In the case of Latin America, where penetration averaged 7.66 % in 2011, broadband price elasticity was −1.88. In the case of the OECD, with an average broadband penetration of 27.48 %, price elasticity was −0.53. This result begins to point out that, as expected, mature markets tend to be more price inelastic. Coincidentally, Lee et al. (2011) found that for OECD countries between 2003 and 2008, elasticity was −1.58 (lower than for Latin America). Confirming the inverse relation between penetration and elasticity, Dutz et al. (2009) analyzed the elasticity of broadband service in the US between 2005 and 2008. They found that in 2005 the elasticity was of −1.53 but declined to −0.69 in 2008, confirming the declining elasticity trend at higher penetration levels.

Table 4.2 presents the summary results of all studies briefly reviewed above.

Considering the level of service adoption at the time each of these studies was completed allows estimating the relationship between fixed broadband penetration and elasticity (see Fig. 4.2).

Table 4.2 Studies on broadband service price elasticity

References	Year	Elasticity
Goolsbee (2006)	USA state level 1998/ 1999	Between −2.8 and −3.5 (penetration between 2 and 3 %)
Rappoport et al. (2002)	USA 2000	−1.491
Dutz et al. (2009)	USA 2005	−1.53
Lee et al. (2011)	OECD 2003/2008	−1.58
Dutz et al. (2009)	USA 2008	−0.69
Cadman and Dineen (2008)	OECD 2007	−0.43
Galperin and Ruzzier (2013)	OECD + LATAM 2011	LATAM: −1.88 OECD: −0.53

Source Compiled by the authors

Fig. 4.2 Correlation between fixed broadband (FBB) penetration and price elasticity. *Note* the *t*-value for the variable of interest is −0.755. *Source* Estimates by the authors based on research literature

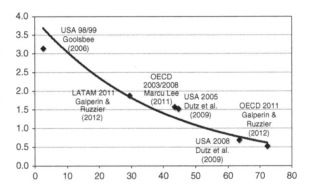

While the elasticity data in Fig. 4.2 is presented in absolute values, the price elasticity coefficient is always negative indicating the indirect relationship between price and demand. Thus, the relationship between both variables indicates that a change in the price level would have a positive impact in the level of penetration of fixed broadband. By relying on the estimates depicted in Fig. 4.2, the effect of a price reduction between 5 and 25 % was estimated for different regions of the world (see Table 4.3).

As indicated in Table 4.3, the price elasticity is higher for the regions with lower levels of penetration. As a result, in regions like Sub Saharan Africa or South Asia, a 25 % price decline could yield an approximate doubling of current penetration levels. The increase in fixed broadband penetration is substantial in other emerging countries as well (see Table 4.4).

Table 4.3 Impact on weighted average penetration level (HH)[a] of fixed broadband of a price reduction

Region	2011 HH penetration (%)	5 % price reduction (%)	10 % price reduction (%)	15 % price reduction (%)	20 % price reduction (%)	25 % price reduction (%)
East Asia and Pacific	38.96	41.75	44.53	47.32	50.11	52.89
Europe and Central Asia	54.95	57.54	60.13	62.73	65.32	67.91
Latin America and Caribbean	28.71	31.39	34.07	36.75	39.43	42.11
MENA	16.73	18.86	21.00	23.13	25.26	27.40
North America	76.36	78.42	80.49	82.56	84.62	86.69
South Asia	4.80	5.64	6.47	7.31	8.14	8.98
Sub Saharan Africa	1.14	1.36	1.58	1.79	2.01	2.23

Source Estimates by the authors based on ITU 2011 data

[a] *HH* refers to households

Estimating the Impact of Broadband Price Declines

Data on broadband price elasticity models was utilized to develop a simple model template for estimating the impact of a reduction in subscription prices on service penetration. Follow these steps:

a. Locate the country in one of the following seven regions (East Asia and Pacific, Europe and Central Asia, Latin America and Caribbean, MENA, North America, South Asia, Sub Saharan Africa)
b. Input total number of households, fixed broadband subscriptions, and average monthly subscription charge

The model will provide the yield in broadband household penetration by price reduction scenarios (5, 10, 15, 20, 25 %). The formula underlying the model and derived from the data in Fig. 6.25 is the following:

$$\text{Expected penetration} = 3.9393e^{-0.026 * \text{Current Penetration Level} * \text{Price reduction}}$$

3.9393: Expected Penetration: penetration resulting from a price reduction
−0.026: Price elasticity coefficient resulting from a decline in pricing of 1 %
Current Penetration: penetration of broadband at time of test
Price reduction: hypothetical price reduction

Region	Country	Households	Fixed broadband subscriptions	% Households	Monthly subscription charge, in USD	5 % price reduction	10 % price reduction
East Asia and Pacific							
Europe and Central Asia							
Latin America and Carib.							
MENA							
North America							
South Asia							
Sub Saharan Africa							

4.2.2.2 Mobile Broadband Price Elasticity

In the section above, the evidence provided showed that fixed broadband service pricing was indirectly linked to penetration levels. Unfortunately, given the recent deployment of mobile broadband, still no substantial evidence exists that links

Table 4.4 Growth of penetration level (per household) of fixed broadband as a consequence of a price reduction

Region	5 % price reduction (%)	10 % price reduction (%)	15 % price reduction (%)	20 % price reduction (%)	25 % price reduction (%)
East Asia and Pacific	7.15	14.31	21.46	28.61	35.76
Europe and Central Asia	4.72	9.44	14.16	18.88	23.60
Latin America and Caribbean	9.34	18.68	28.01	37.35	46.69
MENA	12.75	25.50	38.25	50.99	63.74
North America	2.70	5.41	8.11	10.82	13.52
South Asia	17.38	34.77	52.15	69.54	86.92
Sub Saharan Africa	19.12	38.24	57.36	76.48	95.61

Note For each tariff reduction scenario, the growth on broadband penetration was estimated based on the formula of Fig. 6.26
Source Estimates by the authors based on ITU 2011 data

pricing and penetration.[5] However, existing pricing data for OECD and Latin American countries allows estimating the elasticity for a 3G connection. For this purpose, six models were specified for mobile broadband plans dongles/air cards and smartphones:

- Smartphone plans with 500 MB monthly cap
- Smartphone plans with 1 GB monthly cap
- Smartphone plans with 2 GB monthly cap
- Air card/dongle plans with 1 GB monthly cap
- Smartphone plans with 2 GB monthly cap
- Smartphone plans with 5 GB monthly cap

Data used for estimating elasticity coefficients comprised the following (Table 4.5):

The regressions, included in the appendix, indicate that the price elasticity for air cards and dongles is twice that one for handsets (see Table 4.6).

According to these models, a 10 % price decline in the mobile broadband plan for a smartphone connection would generate a penetration increment between 2.35 and 3.20 %. On the other hand, a 10 % price decline in air cards/dongles plans for personal computer connectivity could yield an approximate increment of penetration level between 6.33 and 6.73 %. In the second situation, the increment would not only come from new broadband subscribers; it would comprise a

---

[5] See the first reported study by Srinuan et al. (2011) on the Swedish market based on 2009 data.

Table 4.5 Variables for the econometric estimation of the mobile broadband price elasticity model

Variables (in logarithm)	Explanation	Source
Mobile broadband penetration	Dependent variable. Measures the level of penetration of mobile broadband (3G and 4G) by country	Wireless Intelligence
Mean of mobile broadband plan prices (handset/air card)	Explanatory variable. Is the average of the prices of the plans available for each plan	Telecom Advisory Services
GDP per capita	Control for GDP as an indicator of standard of living	World Bank
Urbanization rate	Control for level of urban population for each country	World Bank
Households with computer ownership	Control for share of houses with ownership of a computer	World Bank/ ITU
Price of 1 min call on mobile phone off net	Control for price of a call with a mobile phone	ITU
Share of population with coverage for mobile phone	Control for the level of deployment of the mobile network	ITU

Table 4.6 Price elasticity coefficients for different price plans

CAP	Handset	Air cards/ dongles
500 MB	−0.320 (−2.31)[b] 80.70 %	
1 GB	−0.305 (−2.26)[b] 80.60 %	−0.633 (−4.34)[a] 83.27 %
2 GB	−0.245 (−1.85)[c] 81.88 %	−0.667 (−4.68)[a] 83.38 %
5 GB		−0.673 (−4.94)[a] 84.06 %

[a] Significant at 5 %
[b] Significant at 10 %
[c] Significant at 15 %
Source Estimates based on models specified by the author

substitution effect from fixed broadband to mobile broadband. Finally, elasticity coefficients for handsets with higher CAPs tend to be lower than the basic plans because high volume subscribers are less price sensitive, while lower CAP subscribers are primarily email and Facebook users, implicitly low value subscribers.

4.2.3 The Effect of Competition on Broadband Pricing

Having reviewed the importance of service pricing in limiting adoption of broadband, it is relevant to address potential policy initiatives that could yield a reduction in tariffs. As it has been considerably researched, the development of competition is one of the major tools for affecting a reduction in telecommunications service pricing. The theoretical basis of competition is the notion that, in the telecommunications market, multiple operators can compete among each other and generate sufficient benefits for consumers in terms of price reductions, while guaranteeing an appropriate rate of innovation. The following features characterize a telecommunications competition model:

- Existence of multiple operators serving the same market based on their own network.
- Existence of multidimensional competitive dynamics (prices, services and user service quality) among industry players.
- Reduction of retail prices for consumers, and intense competition in product differentiation (dynamic efficiencies), resulting in additional consumer surplus.
- Competitive stimulation for each operator to increase the level of investment in its own network.
- Absence of tacit collusion between operators due to the high rate of innovation and competition based on product differentiation.

A number of countries around the world have already implemented this list of principles resulting in competition for broadband as a model for organizing the industry (see Table 4.7).

The industry structure in these countries not only includes a facilities-based telecommunications operator and one (or more) cable operator, but also a second mobile/landline telecommunications operator and at least one mobile operator competing with the landline operators on an intermodal scale.

A competitive market structure has a positive influence on the reduction of broadband prices. For example, the authors determined that in Latin America, the average monthly price of a basic fixed broadband connection declined from US$ 21.06 in 2010 to US$ 17.46 in 2012 (a reduction of 17 % in 2 years). Table 4.8 presents price reductions in the region across fixed and mobile broadband plans.

As Table 4.8 indicates, broadband prices in Latin America have been declining, albeit at different rates. To understand the importance of competition in driving the price reductions, Katz (2012) completed an analysis of the relationship between competitive intensity (as measured by the Herfindahl Hirschman Index) in the Latin American broadband industry and retail prices (see Fig. 4.3).

The correlation coefficient between broadband competitive intensity and broadband prices is 0.63 %. The Fig. 4.3 also indicates that, even at moderate competitive levels (HHI < 4000), prices tend to cluster at the lower levels within

Table 4.7 National market shares (2011)[a]

	United States	Netherlands	Korea	Chile	Canada
Wireline	• Telco 1 (38.4 %) • Telco 2 (25.4 %) • Cable (11.42 %)	• Telco 1 (55 %) • Cable (23.6 %)	• Telco 1 (77.0 %) • Telco 2 (13.0 %) • Telco 3 (10.0 %)	• Telco 1 (56.5 %) • Cable (18.0 %) • Telco 2 (5.4 %) • Telco 3 (5.5 %)	• Telco 1 (36.3 %) • Cable (6.0 %) • Telco 2 (21.0 %)
Wireless	• Telco 1 (30.1 %) • Telco 2 (32.1 %) • Telco 3 (11.7 %) • Telco 4 (16.6 %)	• Telco 1 (47.2 %) • Telco 2 (28.5 %) • Telco 3 (24.2 %)	• Telco 1 (31.4 %) • Telco 2 (50.7 %) • Telco 3 (17.9 %)	• Telco 1 (56.5 %) • Telco 2 (37.5 %) • Telco 3 (20.7 %)	• Telco 1 (30.0 %) • Cable (37.0 %) • Telco 2 (28.6 %)
Broadband	• Telco 1 (20.3 %) • Telco 2 (11.8 %) • Cable (35.5 %)	• Telco 1 (41.7 %) • Cable (36.6 %)	• Telco 1 (42.7 %) • Telco 2 (23.4 %) • Telco 3 (15.6 %)	• Telco 1 (44.3 %) • Cable (38.5 %) • Telco 2 (1.2 %) • Telco 3 (6.4 %)	• Telco 1 (20.3 %) • Cable (16.1 %) • Telco 2 (11.1 %)
Content distribution	• Cable (39.9 %) • Telco 1 (4.49 %) • Telco 2 (3.25 %)	• Cable (68.7 %) • Telco 1 (14 %)	• Cable (85 %) • Telco 1 (7.3 %) • Telco 2 (4.9 %)	• Cable (51.7 %) • Telco 1 (17.0 %) • Telco 3 (17.5 %)	• Telco 1 (17.1 %) • Cable (20.0 %) • Telco 2 (1.7 %)
Companies	• Telco 1: ATT • Telco 2: Verizon • Telco 3: T-Mobile • Telco 4: Sprint • Cable: Comcast, Cablevision and TWC	• Telco 1: KPN • Telco 2: Vodafone • Telco 3: T-Mobile • Cable: UPC, Ziggo	• Telco 1: KT • Telco 2: SK/Hanaro • Telco 3: LG	• Telco 1: Telefónica • Telco 2: ENTEL • Telco 3: Telmex/Claro • Cable: VTR	• Telco 1: Bell Canada • Telco 2: Telus • Cable: Rogers

[a] Number in brackets depicts market share

Sources National regulatory agencies; company reports

Table 4.8 Latin America: broadband average monthly subscription prices (2010–2012)[a] (in USD)

	Plan	2Q2010	2Q2011	2Q2012	Decline
Fixed broadband	Basic plan with 2 GB cap	$ 21.06	$ 18.71	$ 17.46	(17.09)
	Least expensive 2.5 Mbps and 6 GB cap	$ 77.97	$ 53.05	$ 44.14	(43.38)
	Least expensive 6.0 Mbps and 6 GB cap	$ 89.73	$ 78.48	$ 82.70	(7.83)
Mobile broadband	Least expensive PC plan with 1 GB cap	$ 19.59	$ 17.60	$ 14.39	(26.54)
	Least expensive smartphone plan with 250 MB cap	$ 17.68	$ 12.79	$ 9.24	(47.74)
	Least expensive smartphone plan with 1 GB cap	$ 23.07	$ 18.71	$ 16.33	(29.21)

[a] Includes Argentina, Bolivia, Brazil, Chile, Colombia, Costa Rica, Ecuador, El Salvador, Mexico, Nicaragua, Panama, Paraguay, Peru, Dominican Republic, Uruguay, and Venezuela
Source Katz (2012)

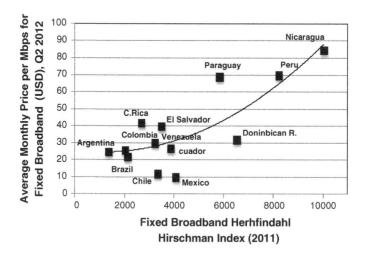

Fig. 4.3 Latin America: fixed broadband competition and price per Mbps. *Source* Katz (2012)

the region. A similar analysis was conducted but in this case, the HHI index was calculated by adding the market shares of fixed and mobile providers.[6]

Figure 4.4 depicts a directional correlation between the index of competitive intensity in the broadband market and prices of fixed broadband offerings at the 6 Mbps download speed level. This relationship indicates an embryonic convergence between the fixed and mobile broadband markets that is having a positive

---

[6] In the case of convergent players, present in both fixed and mobile markets, the shares in both markets were added before calculating the HHI index.

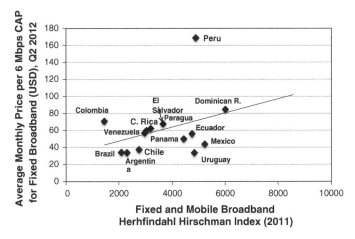

Fig. 4.4 Latin America: fixed and mobile broadband competition and price per 6 Mbps of fixed broadband. *Source* Katz (2012)

influence in the price of fixed broadband. In this context, one can point out at the positive influence that mobile broadband launch has had on overall service prices.

As a side note, it is important to mention that it is common in the developed world and some emerging countries that telecom operators, in the context of competitive pressures, offer first time customers discounts and free promotional offers covering activation fee, first-year monthly subscriptions, etc. This marketing effort, resulting from competitive incentives prevalent in the market, has also a positive contributing impact on total cost of ownership. This trend is having already a significant contribution to reducing the universe of non-broadband adopters, as it is analyzed in the next section.

4.2.4 The Role of Mobile Broadband in Lowering the Affordability Barrier at the Bottom of the Pyramid

The bottom of the sociodemographic pyramid refers to those individuals and households with low income.[7] Although not one single definition of this segment exists, this social group has been conceptualized in four ways:

- Households with a daily per capita income lower than US$ 2.50 (Shah 2013)
- Households with an income that places them below the poverty line at the national level (InfoDev 2012)

---

[7] The following section reprises portions of Katz and Callorda (2013). *Mobile broadband at the Bottom of the Pyramid in Latin America*. London: GSMA.

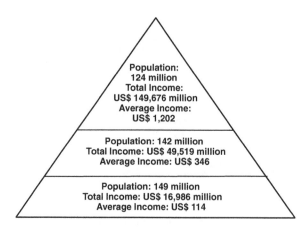

Fig. 4.5 Latin America: composition of the income pyramid (2012). *Note* The *base* of the pyramid corresponds to the 3 lowest income deciles, the *top* of the pyramid corresponds to the three deciles and average of the highest incomes, and the *middle* of the pyramid corresponds to the rest of the population. *Sources* Argentina (INDEC 2012); Brazil (IBGE 2012; Euromonitor); Colombia (DANE 2012; Euromonitor); Ecuador (INEC 2012); México (INEGI 2012; Euromonitor)

- Households located at the bottom three income deciles in a given country (NsrinJazani 2011)
- Individuals with a per capita annual income lower than US$ 1,500 in purchasing parity (Prahalad 2004; Prahalad 2010).

Despite the differences in the cutoff point that separates the bottom from the rest of the pyramid, all four definitions coincide in determining that this segment represents an important portion of the population in developing countries. For example, in Latin American countries, the three lowest income deciles cover those individuals whose average daily incomes fall below the US$ 2.50 cutoff point.[8] Figure 4.5 presents the monthly per capita income by decile for five Latin American countries.

The three bottom income deciles of Latin American nations exhibit substantial differences. For example, the profiles of average income of Argentina and Mexico are relatively similar (first decile, US$ 96 and US$ 72, respectively; third decile, US$ 251 for both countries). On the other hand, while the average income of the first decile in Brazil and Colombia is relatively similar (US$ 39 and US$ 34, respectively), the income for the third decile of the two countries is significantly different (US$ 161 and US$ 89, respectively). The number of individuals and homes that comprise the base of the pyramid also varies by country. Without a

---

[8] The utilization of the three bottom deciles allows the analysis to overcome the heterogeneity among countries in their determination of the poverty line. For example, in Argentina, the poverty level is 5 % while in Mexico it is 40 %.

Table 4.9 Latin America: population and households at the bottom of the pyramid (2012)

Country	Population	Households
Argentina	16,937,000	3,933,000
Brazil	74,970,000	15,300,000
Colombia	16,335,000	2,970,000
Ecuador	5,058,000	1,140,000
México	35,424,000	9,480,000

Sources Argentina (INDEC 2012); Brazil (IBGE 2012); Colombia (DANE 2012); Ecuador (INEC 2012); México (INEGI 2012)

doubt, this number determines the magnitude of the digital demand gap challenge (see Table 4.9).

As the sum of the statistics in Table 4.9 shows the bottom of the Latin American sociodemographic pyramid in these five countries alone represents 136 million people and 32 million homes. By virtue of the fact that their average monthly income is US$ 114 by person, their capacity to afford broadband access is fundamentally limited.

Mobile broadband has been able to provide affordable service to households at the bottom of the pyramid in several ways. First, as mentioned above, driven by competition in the wireless market, mobile broadband prices, both in personal computer connectivity (through USB modems) plans and in data plans for smartphones, have been significantly reduced in recent years. Second, mobile broadband offers pricing flexibility that allows consumers to purchase services based on what they can afford (by day, by download volume, or by type of Internet service been accessed). Third, mobile access to the Internet through smartphones overcomes other barriers to broadband adoption at the bottom of the pyramid (such as, for example, the cost of purchasing a personal computer, limited digital literacy, or lack of access to electricity). Each of these three aspects of the value proposition of mobile broadband will be analyzed in more detail in turn.

4.2.4.1   Reduction of the Price of Mobile Broadband

The mobile broadband market experienced major growth in recent years. This process increased the incentives for operator entry and expansion (building or increasing the coverage of 3G networks or launching LTE services). Figure 4.6 shows the evolution of the competitive intensity in terms of the Herfindahl Hirschman Index (HHI)[9] between the first quarter of 2010 and the first quarter of 2013 for five Latin American countries.

---

[9] The HHI index measures industry concentration or fragmentation, and is calculating by adding the square of the market shares of operators serving a particular market. As such, it serves to indicate the intensity of competition. A lower index points to higher industry fragmentation and,

Fig. 4.6 Latin America: Herfindahl–Hirschman index of competitive intensity in the mobile broadband market (1Q10–1Q13). *Source* Katz and Callorda (2013) estimates based on data from GSMA Intelligence

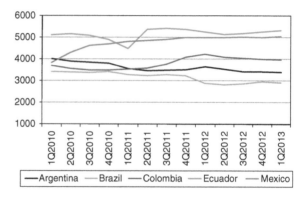

In the period covered in this analysis, the average value of the HHI index for the mobile broadband market in Latin America fell to 4,064 from the previous level of 5,419, an average annual reduction of 9.15 %. This finding demonstrates that the intensity of competition in the mobile broadband market within the region is increasing as a whole. In some countries, like Brazil (HHI: 2,911), and Argentina (HHI: 3,406), competition in the mobile broadband market is intense. While the competitive intensity in these countries has increased in recent years, it has declined in Mexico (and to some extent in Colombia) while remaining stable in Ecuador.

As expected, the greater the competitive intensity, the more pronounced the price reduction. As an example, in Costa Rica the mobile broadband plan with at least 1 GB of monthly download limit went from US\$ 20.34 in 2010, with the presence of just one player in the market, to a monthly price of US\$ 13.56 in 2011 (months before the entry of two competitors) and later to US\$ 7.94. In 2012, with the presence of three operators, it finally fell to US\$ 5.05.

Figure 4.7 shows the price of the most economic mobile broadband plans that offer a monthly download capacity greater than or equal to 1 GB for the years 2010 through 2013 in Latin American countries.

As shown in Fig. 4.7, the average price of this type of data plan in Latin America has consistently fallen in a significant way. In 2010, the average price of the least expensive plan was US\$ 19.59. By 2011 it had dropped to US\$ 17.60 and by 2013 it reached US\$ 15.60 (an annual decline of −7.31 %). In this regard, it should be noted that the trend toward price reduction is most notable between 2010 and 2012; by 2013, pricing stayed relatively constant in the majority of the countries. The slowdown in price reduction is largely a consequence of the plan's success. Mass adoption of the offering resulted in a spike in demand for mobile network capacity and, consequently, its saturation. This in turn led operators to

(Footnote 9 continued)
therefore, more competition, while a higher index points to higher concentration, and increase in market power.

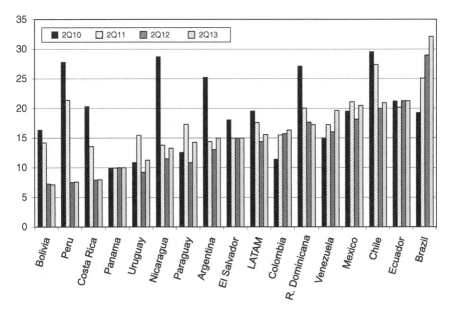

Fig. 4.7 Latin America: most economic mobile broadband plan for computers with at least 1 GB of download volume cap (US$). *Source* Katz and Callorda (2013) based on data from Galperin (2012) and TAS research

moderate their commercial aggressiveness in their sales push of least expensive data products. This effect has spectrum availability implications to the extent that if operators had access to additional frequency spectrum, they would be able to continue to aggressively marketing these products, thus maximizing penetration. Furthermore, network capacity could be improved if operators had less restrictions regarding infrastructure deployment.

On the other hand, it is interesting to note that the most economic plans of mobile broadband (for computer connectivity using a USB modem) are found in Bolivia and Peru, countries that have the most onerous plans for fixed broadband. In particular, it is important to highlight the case of Bolivia where, in the first quarter of 2010, there was only one mobile broadband provider, which offered its least expensive plan at US$ 16.38. With the entrance of VIVA (Nuevatel) to the market, the price of the cheapest plan dropped significantly, reaching US$ 7.13 per month in the second quarter of 2013. This decline in prices generated a cross-elastic substitution effect that resulted in a fall in fixed broadband penetration during the studied period, and a dramatic increase in the penetration of mobile broadband connections. The example of Bolivia demonstrates that a disruptive decrease in the price of mobile broadband plans can generate a substitution effect (away from fixed broadband). This, combined with the ability to extend deployment of the 3G network (in relation to the deployment of ADSL or cable modem), allows the technology to serve regions that otherwise would not be able to access the service, thus satisfying the needs of the most vulnerable social sectors.

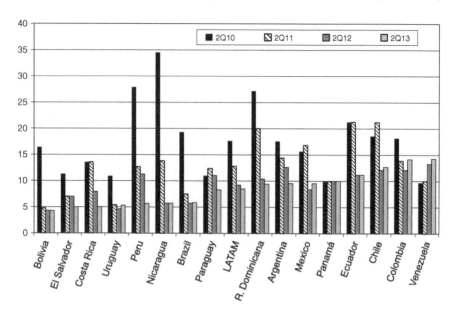

Fig. 4.8 Latin America: most economic broadband plan for smartphones with at least 250 MB download cap (US$). *Source* Katz and Callorda (2013) based on data from Galperin (2012) and TAS survey

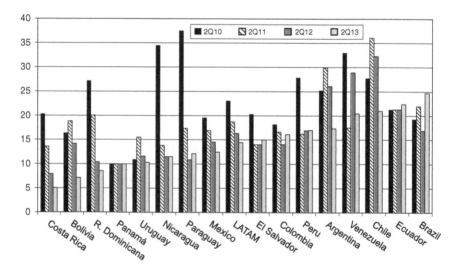

Fig. 4.9 Latin America: most economic mobile broadband plans for smartphones with a minimum download cap of 1 GB (US$). *Source* Katz and Callorda (2013) based on data from Galperin (2012) and Telecom Advisory Services survey

Beyond plans offering computer connectivity through USB modems, mobile broadband offers plans to access the Internet through mobile devices (smartphones or tablets). The prices for this type of plan generally start with a minimum content download cap. Figure 4.8 presents each country's most economic plans for accessing mobile broadband from a smartphone or tablet with a download limit of at least 250 MB per month.

According to Fig. 4.8 in all of the countries in the region, basic smartphone data plans are offered for less than US$ 15 per month (from as low as US$ 4.28 in Bolivia to US$ 14.29 in Venezuela). Further, it is important to highlight the declining price of these plans between 2010 and 2013 in, with the exception of Venezuela, all of the countries. The second quarter of 2010 showed an average monthly price of US$ 17.68, which fell to US$ 12.79 in 2011 and, finally, to US$ 8.33 in 2013 (a 52 % reduction in 3 years). These mobile broadband plans with more affordable prices relative to fixed broadband plans contribute to an increase in broadband adoption amongst mobile phone users. In this way, mobile broadband not only complements fixed-line services—offering access to email, social networks, or web browsing outside the home—but also serves as a gateway to Internet access for the population in the bottom of the pyramid who otherwise cannot afford the monthly cost of fixed access.

Because the download cap of 250 MB per month only allows for a limited number of functions (like access to social networks or email), Fig. 4.9 presents the evolution of low-cost plan pricing for mobile devices with a monthly download capability of at least 1 GB (the amount needed for smartphone users to regularly access most Internet content, except for video-streaming).

Pricing of this type of plans also fell sharply in recent years[10] from an average US$ 23.07 per month in 2010, to US$ 18.71 in 2011, and to US$ 14.44 in 2013 (a decline of 37 % in 3 years).

To conclude, since competitive intensity is inversely related to price reductions, high competition in the mobile broadband market played an important role in price reductions.[11] The dramatic decline in mobile broadband pricing raises the possibility for the service to meet the needs of population at the bottom of the pyramid. As an example, the basic mobile broadband plan with USB modems for computers in Argentina raised the level of broadband affordability through the third income decile of the population.

The bottom of the Argentine sociodemographic pyramid is comprised of 3,933,000 homes in the lowest three deciles exhibiting a range of average monthly household income between A$ 2,431 and A$ 5,118.[12] Considering this income level, even under the conditions where fixed broadband pricing has not increased

---

[10] The only exceptions are Panama and Ecuador, where it remained virtually constant, and Brazil, where the price of this product increased.

[11] The correlation between the decrease in the competition index (HHI) and the decline of mobile broadband prices in 2011 was 40 %.

[12] This equates to 4,389,000 individuals with a monthly income between A$ 510 and A$ 1,329.

		DECILE 1	DECILE 2	DECILE 3	DECILE 4	DECILE 5	DECILE 6	DECILE 7	DECILE 8	DECILE 9	DECILE 10
Total Household Income		$ 2.431	$ 4.088	$ 5.118	$ 5.522	$ 5.796	$ 7.026	$ 8.561	$ 9.489	$ 10.820	$ 16.724
Basic Fixed Broadband Plan	Monthly Subscription	A$ 123.33									
	Percentage of monthly subscription	5.07%	3.02%	2.41%	2.23%	2.13%	1.76%	1.44%	1.30%	1.14%	0.74%
Medium Fixed Briadband Plan	Monthly Subscription	A$ 133.33									
	Percentage of monthly subscription	5.48%	3.26%	2.61%	2.41%	2.30%	1.90%	1.56%	1.41%	1.23%	0.80%

AFFORDABILITY THRESHOLD: The monthly subscription cannot exceed 2% of the household total income (considering other communication expenditures such as wireless subscription)

Fig. 4.10 Argentina: affordability of fixed broadband plans (2013). *Source* Katz and Callorda (2013)

		DECILE 1	DECILE 2	DECILE 3	DECILE 4	DECILE 5	DECILE 6	DECILE 7	DECILE 8	DECILE 9	DECILE 10
Total Household Income		$ 2.431	$ 4.088	$ 5.118	$ 5.522	$ 5.796	$ 7.026	$ 8.561	$ 9.489	$ 10.820	$ 16.724
Basic Mobile Broadband Plan	Monthly Subscription	Cost of Service (A$90.00)+ (Cost of Modem (A$ 139)/18 months)									
	Percentage of monthly subscription	4.02%	2.39%	1.91%	1.77%	1.69%	1.39%	1.14%	1.03%	0.90%	0.58%

Additional deciles

Fig. 4.11 Argentina: affordability of basic plan of mobile broadband for PC. *Source* Katz and Callorda (2013)

commensurately to salaries, fixed broadband plans, at prices ranging between A$ 123.33 and A$ 133.33, is beyond the economic reach of these households. This premise is based on the assumption that broadband is not affordable if it exceeds 2 % of total household income (considering that, of the 5 % of income spent in communications and entertainment, 3 % must be assigned to other services like cellular telephony)[13] (See Fig. 4.10).

According to Fig. 4.10, considering that the monthly price of a basic fixed broadband plan starts at A$ 123.33 and the average home does not spend more than 2 % on a broadband subscription, this product is not affordable to homes

---

[13] The International Broadband Commission establishes a limit of 5 % of income for communication and entertainment costs.

Table 4.10 Latin America: examples of daily charge plans (2013)

	Operator	Terminal	Daily price (in local currency)	Daily price (US$)
Argentina	Claro	Dongle	A$ 5.00	0.96
		Smartphone	A$ 1.00	0.19
Bolivia	Tigo	Smartphone	B$ 2.00	0.29
Brazil	Vivo	Dongle	R$ 2.99	1.48
	TIM	Smartphone	R$ 0.50	0.25
Chile	Claro	Dongle	C$ 1,886.00	3.98
	Entel	Smartphone	C$ 250.00	0.53
Colombia	Movistar	Dongle/Smartphone	C$ 2,900.00	1.56
México	Movistar	Smartphone	$M 19.00	1.58
		Dongle	$M 15.00	1.24
Uruguay	Claro	Smartphone	$U 10.00	0.52
		Dongle	$U 20.00	1.57

Source Katz and Callorda (2013)

whose income is below the average of the sixth decile. In this context, mobile broadband can help address this market failure. For example, the basic mobile broadband plan for USB modems for PCs brings the level of affordability down to the third income decile. For this, the sum of the monthly service subscription (A$ 90.00) and the cost of the modem (A$ 139) divided by 18 months[14] is taken. Under this premise, the total cost of ownership does not exceed 2 % of the income of households through the third income decile (See Fig. 4.11).

4.2.4.2 Pricing Flexibility

Beyond the aggregated decline of mobile broadband prices, operators have introduced an extensive array of plans that support a variety of broadband use cases. Through a daily charge plan, for instance, the subscriber acquires the right to access the Internet on a daily basis at a price paid each month as a function of the number of days purchased. This allows the user to control the amount spent per month based on what is economically feasible. Table 4.10 presents examples of this type of plan offered in the Latin American region.

Beyond daily use mobile broadband access, customers can also purchase plans based on download volume cap (20 or 100 MB per month), which also promotes flexibility. This option allows the subscriber to have permanent access to the Internet while adapting the consumption pattern to the type of content accessed. For example, if the subscriber uses the Internet primarily to access social networks (like Facebook and Twitter), the 20 Mbps package is adequate. Obviously, this alternative substantially restricts access to "heavier" content, like video clips and YouTube. However, it is an initial option for those members of the population who

---

[14] Which assumes a replacement rate of 18 months.

Table 4.11 Latin America: examples of plans with limited download volumes (2013)

Country	Operator	Download limit (MB)	Daily price (local currency)	Daily price (US$)
Argentina	Claro	100	A$ 55.00	10.53
Brazil	Oi	100	B$ 9.90	4.90
Bolivia	Viva	7	B$1.00	0.19
Chile	Entel	50	C$ 500	1.06
Colombia	Claro	100	C$ 2,890.00	1.56
Ecuador	Claro	20	US$ 1.11	1.11
Mexico	Telcel	50	M$ 49.00	4.07
Paraguay	Personal	100	P$ 100,000	2.43

Source Katz and Callorda (2013)

Table 4.12 Latin America: pricing of restricted use plans

Country	Colombia	Nicaragua	Mexico
Email package	$ 9,900 US$ 5.25	N/D	N/D
Chat package	$ 9,900 US$ 5.25	N/D	N/D
Email + Chat package	$ 15,900 US$ 8,43	N/D	$ 100 US$ 8,33
Social network package	N/D	US$ 10.99	$ 80 US$ 6.66
Social network + email + chat package	$ 19,900 US$ 10.55	US$ 13.99	N/D
Navigation package	$ 24,900 US$ 13.20	US$ 24.99	N/D

Sources Katz and Callorda (2013) based on a review of operator web pages (in all cases, the operator that offers these packages is Movistar)

do not otherwise have access and goes beyond the basic access offered in a public center. Capped access download plans are very popular in Latin America, as seen in the examples below (See Table 4.11).

Finally, another mechanism used by regional operators to increase service penetration is to offer differentiated prices based on the user's type of broadband use. In this way the plans that only include email (Gmail) or chat (MSN, Talk, and Yahoo Messenger) are more affordable. A variety of those plans are the ones that offer unlimited use of that particular access mode during the month. Along these lines, the price will increase by an additional 50 % when purchasing unlimited use of social networking (Facebook and Twitter); finally, the most expensive plan includes unlimited use of YouTube and Google. The following table shows the rates for these plans in Latin America (Table 4.12).

The following examples of use cases show how flexible pricing can offer a solution to the demand gap at the bottom of the sociodemographic pyramid (see Table 4.13).

Table 4.13 Use cases where mobile broadband can contribute to addressing the gap at the bottom of the pyramid

Situation	Complication	Solution
User has purchased a used computer at an affordable price	Difficulty with monthly payments for fixed broadband	Purchase a USB modem at a low price (to control download volume or use frequency)
User has purchased a subsidized smartphone, either as part of a handset replacement or through payment installments	User wants to access the Internet but cannot afford a computer or tablet	Purchase a plan with capped download volume (sometimes operators include this offer as part of a subsidized package)
User has purchased a computer at an affordable price	The limited coverage of fixed broadband does not allow for affordable access	Purchase of a USB modem at a low price (for control of download volume or use limited to a reduced number of days)

		DECILE 1	DECILE 2	DECILE 3	DECILE 4	DECILE 5	DECILE 6	DECILE 7	DECILE 8	DECILE 9	DECILE 10
Total Household Income		$ 456	$ 989	$ 1.448	$ 1.915	$ 2.435	$ 3.053	$ 3.847	$ 4.986	$ 7.000	$ 17.102
Basic Fixed Broadband Plan	Monthly Subscription	R$ 29.80 (National Broadband Plan)									
	Percentage of monthly subscription	6.54%	3.01%	2.06%	1.56%	1.22%	0.98%	0.77%	0.60%	0.43%	0.17%
Medium Fixed Broadband Plan	Monthly Subscription	R$ 59.90									
	Percentage of monthly subscription	13.14%	6.06%	4.14%	3.13%	2.46%	1.96%	1.56%	1.20%	0.86%	0.35%

AFFORDABILITY THRESHOLD: The monthly subscription cannot exceed 2% of the household total income (considering other communication expenditures such as wireless subscription)

Fig. 4.12 Brazil: affordability of "popular" broadband plan. *Source* Katz and Callorda (2013)

These examples of "use cases" are possible in many emerging countries. For example, the daily computer connection plans make Internet access significantly more affordable for the population at the bottom of the pyramid in their respective countries. For example, in Brazil, considering that the average individual income at the bottom of the pyramid is between $R 456 and $R 1.448, fixed broadband (even the "popular broadband" plan offered as a result of the National Broadband Plan) is not affordable beyond the fourth decile of the population (see Fig. 4.12).

In this sense, while the "social" fixed broadband offer addresses the affordability obstacle for the middle classes, it does not solve the gap at the bottom of the pyramid. In this context, mobile broadband plans with utilization caps (in terms of the number of days per month) are able to address this failure. For example, the

		DECILE 1	DECILE 2	DECILE 3	DECILE 4	DECILE 5	DECILE 6	DECILE 7	DECILE 8	DECILE 9	DECILE 10
Total Household Income		$ 456	$ 989	$ 1.448	$ 1.915	$ 2.435	$ 3.053	$ 3.847	$ 4.986	$ 7.000	$ 17.102
Daily Mobile Broadband Plan	Cost	150 MB por 2 days (R$ 2.99) * 4 + (Cost of Modem (R$ 119)/18 months)									
	Percentage of monthly subscription	4.07%	1.88%	1.28%	0.97%	0.76%	0.61%	0.48%	0.37%	0.27%	0.11%

Additional deciles

Fig. 4.13 Brazil: affordability of the daily offer of mobile broadband. *Source* Katz and Callorda (2013)

daily mobile broadband service used 8 days a month in Brazil is able to shift the affordability barrier through the second decile (see Fig. 4.13).

The same effect is observed in the plans of limited consumption, like the offers in Colombia, Argentina, Mexico, and Ecuador (see case studies).

4.2.4.3 Smartphone Contribution

Beyond the price of the service, the relative low cost of the mobile broadband device (smartphone) involves other benefits. In the first place, smartphones serve as a terminal for accessing the Internet whose purchase price is much lower than that of a computer. In this sense, these terminals contribute to reduce the demand gap of the most vulnerable economic sectors.

On the other hand, mobile broadband has characteristics that allow for the provision of connectivity to individuals who otherwise would not be able to purchase it, because of limited digital literacy. For example, mobile broadband does not require significant abilities compared to those skills necessary to operate a computer, as in the case of fixed broadband. This would solve certain barriers posed by lack of digital literacy.

Further, for the individuals in the lowest income decile, lack of electricity can restrict computer use. For example, in Ecuador, 7.80 % of homes do not have electricity. In Colombia, the percentage of homes without electricity is 6.40 %. In this situation, the smartphone, which can be charged outside the home, offers access even for homes that do not have electricity.

The broadband adoption gap in emerging countries is very large. While many public policy initiatives have contributed—in conjunction with a decline in prices—to the increase in penetration of fixed broadband, the effect has been concentrated primarily in the middle of the sociodemographic pyramid. To attack the broadband demand gap at the bottom of the pyramid, that is to say the sectors in the most need, it is necessary to appeal to new strategies that go beyond direct state intervention. Indirect mechanisms that stimulate private investment and competition could be even more fruitful. Mobile broadband, in terms of products that provide connectivity to personal computers and smartphone Internet access plans, represents a solution to this social problem. In the first place, due to the intense competition within the mobile broadband market, prices have declined significantly, thus increasing the affordability of the product. Secondly, the introduction of capped offers, both in temporary terms (days of access) and monthly volumes (MB per month), allow economically vulnerable users to regulate their consumption of and access to the technology, however limited it may be. Finally, Internet access through smartphones can secondarily resolve the gap in the poorest sectors, whose lack of digital literacy prevents them from using a computer.

The importance of these effects within the bottom of the pyramid highlights the need to provide the mobile industry with the necessary inputs to maximize the supply of mobile broadband. In particular, spectrum availability will expand the supply of computer connectivity services, which will result in a reduction of prices and increased service availability. Furthermore, a reduction of taxes to be born by the consumer could have an additional contribution to price declines, which would result in a positive impact on service adoption. Finally, a highly restrictive network neutrality policy could lessen the positive contribution of "variable" mobile broadband effects. In other words, operators should have the freedom to develop mobile broadband offers, with the normal limitations imposed by consumer protection regulations, such as full information on product limitations.

4.2.5 Policy Initiatives Aimed at Reducing the Cost of Broadband Service

Beyond the competitive stimuli, the reduction of broadband service prices can be achieved through a number of targeted public policy initiatives. These initiatives are generally implemented with the objective of achieving universal broadband adoption. The underlying rationale for these policies is that, beyond a competition model, government policies should be implemented to further price reductions of broadband in order to make it accessible to segments of the population affected by limited affordability.

This section will examine four policy options. The first one relies on state-owned telecommunications operators to offer, under their public service imperative, a low-priced broadband service. Obviously, this option is only viable in those countries that have not completely privatized their telecommunications industry. The second option entails a negotiation between the government and private operators for them to offer a low-priced broadband service targeted for disadvantaged segments of the population. The third option is also an agreement between the government and private sector broadband providers to offer low-priced services, but in this case limited to institutions (such as schools, libraries, or health clinics). The fourth option comprises offering free Internet access through Wi-Fi services located in public areas, such as squares, libraries, and transportation hubs.

4.2.5.1 Launch by a Publicly Owned Service Provider

Under this option, a state-owned broadband provider assumes responsibility, as a public service entity, for providing a low-price broadband service. The advantage of this option is that, in addition to fulfilling the objective of tackling the economic barrier, the offering can act as an incentive for other private operators to launch their own more affordable service. Services under this option range from a 256 Kbps line offered for free to existing wireline customers (Uruguay) to a prepaid broadband plan (Venezuela).

Case Study 4.1: Antel (Uruguay)

In May 2011, government-owned telco Antel launched its "Servicio Universal Hogares"—or "Internet for All"—plan, aiming to bring Internet access to every home in Uruguay. For a one-time payment of US$30—the cost of a modem—all fixed line phone customers qualified for free ADSL service. The package offered a basic connection of 256 Kbps and targeted the low-income segment to which the price of broadband represented a barrier to connectivity. At the time, homes and businesses with basic Internet connections paid approximately US$ 150 monthly. Similarly, the Uruguayan government also planned to reach schools and educational institutions with Fiber-to-the-Home (FTTH) technology.

In June 2011, Antel announced plans to connect more than 80,000 Uruguayan households with FTTH by the end of the year. This project initially targeted higher income, urban areas but incorporated plans to reach the lower socioeconomic groups. The rollout incorporated US$100 million investment and a partnership with the Chinese technology firm ZTE. Described as "the most ambitious broadband effort in Latin America," the FTTH project as well as the opening of the Bicentenario submarine cable in early 2012 increased broadband access, speed, and service quality.

The December 2011 launch of its commercial LTE services allowed the telco to offer broadband connections to those regions not yet impacted by the FTTH rollout as well as those customers who could not afford the connectivity costs of fixed Internet.Antel offered customers two package plans from which to choose. By signing a 2-year contract, customers could pay US$ 90 per month for 30 GB. For US$ 76 per month plus an additional $6 in modem rental fees, customers could access 15 GB through a 15-day auto-renew contract.

In time, the Universal Hogares plan expanded, bringing customers faster speeds for lower prices. As of February 2014, the telco offers the following extensions beyond the fixed wireless plan that comes with 1 GB per month at no charge:

Price	Performance (MB)	Details
US$ 2.20	256	30 calendar days from date of purchase
US$ 4.50	512	30 calendar days from date of purchase
US$ 9.00	1024	60 calendar days from date of purchase

Sources

Budde, Paul. "Uruguay—Telecoms, Mobile, Broadband and Forecasts." *Market Research*. N.p., 25 Nov. 2012. Web. http://www.marketresearch. com/Paul-Budde-Communication-Pty-Ltd-v1533/Uruguay-Telecoms-Mobile-Broadband-Forecasts-7256999/.

"Broadband Internet Access Worldwide." *Encyclopedia*. NationMaster, 2006. Web. http://www.nationmaster.com/encyclopedia/Broadband-Internet-access-worldwide.

Prescott, Roberta. "Uruguay's Antel Eyes Mobile Broadband Opportunities with LTE." *RCR Wireless News Americas*. N.p., 20 Apr. 2012. Web. http://www.rcrwireless.com/americas/20120420/carriers/uruguays-antel-eyes-mobile-broadband-opportunities-when-launching-lte/.

"Universal Hogares Rural." *Antel*. N.p., n.d. Web. 27 Feb. 2014. https:// www.antel.com.uy/antel/personas-y-hogares/internet/planes/internet-rural/ universal-hogares.

Case Sutdy 4.2: Plan ABA de CANTV (Venezuela)

In May 2008, state-controlled incumbent fixed-line operator CANTV launched its prepaid broadband Internet access plan known as Plan ABA, or "ABA for Todos." The plan targeted Venezuela's lower-income population and those citizens who did not already have the technology, offering "social rates" to make broadband more affordable and accessible. The plan not only served to increase subscribers for CANTV, but to improve the provider's image as "a company belonging to the state of Venezuela" with the goal of reducing the digital divide. The operator began focusing more of its attention on the provision of more affordable basic services following its 2007 renationalization.

Customers can pay for and manage their broadband usage through CANTV's prepaid calling card, known as the Un1ca card. Launched in 2001, the card offers access to fixed, wireless, and public telephony and Internet services without the obligation of upfront payments or commitments. The card was credited with providing Venezuela citizens with one-stop access to a wide range of services and products, acting as a "communications passport."

Basic services—which included a 256 kpbs connection and a 100 Mb download limit-cost US$ 9.31 monthly, but users could increase capacity for an additional US\$ 0.08 per Mb. Users were also supposed to pay the US\$ 30 Aba subscription fee, which included a modem, but the operator waived the fee. The plan is aimed at prepaid customers; postpaid broadband customers will also require an Un1ca card and must purchase or rent their own modems in order to use the service.

Today, ABA offers a 1.5 Mbps connection speed for \$22.81 per month. In April 2013, the president of CANTV announced a 4 Mbps plan for \$79.20, which equates to nearly 25 % of the minimum wage.

Sources

"Cantv Launches Prepaid Broadband." *Telecom.* Business News Americas, 6 May 2008. Web. http://www.bnamericas.com/news/telecommunications/Cantv_launches_prepaid_broadband.

Annual Report 2001. Rep. CANTV, 2001. Web. http://www.cantv.com. ve/Portales/Cantv/data/InfAnual2001CantvENGLISH.pdf.

"Venezuela." *Freedom House.* N.p., 2013. Web. http://www.freedom house.org/report/freedom-net/2013/venezuela#.UwjQxs0jtFA.

4.2.5.2 Agreement Reached by Private Operators

In this case, government policy makers negotiate with private broadband providers the offering of a low-priced plan. This can be achieved in the context of the formulation of a national broadband plan. Such has been the case of the Brazilian National Broadband Plan referred above in Fig. 4.12, which triggered a negotiation leading to the launch of the "Banda Larga Popular," offered by several operators.

Another option to reach such an agreement could be to attach the offering of a low-priced plan as a *sine qua non* condition for providing regulatory approval of an incumbent plan. Such was the case in the United States, where the government determined that Comcast should offer a low-priced broadband service if it were to receive approval for acquiring NBC Universal. This triggered a process that led all other major cable TV operators to join in the initiative.

A slight variance of this option entails a move by an incumbent wireline operator to offer a low priced plan and create good will in order to preempt a threatening government regulatory move.

Case Study 4.3: Internet Essentials (United States)

In September 2011, cable TV operator Comcast launched its "Internet Essentials" plan to offer broadband to as many as 2.5 million low-income families for a monthly rate of US\$ 9.99. The plan came as part of the approval process in its acquisition of the media and entertainment company,

NBC Universal. Beyond the 1.5 Mbps Internet connection, eligible customers qualified for $150 refurbished computers with software donated by Microsoft. Comcast will also offer digital literacy training to these users free of charge.

To qualify for the plan, households must (a) not yet have a broadband connection and (b) have a child enrolled in a school lunch program. The US$ 9.99 monthly rate lasts for 2 years, at which point customers will likely have the option to renew at a higher—but still discounted—price. Because the US$ 9.99 covers the companies' overhead costs, providers will likely not experience a significant loss in earnings nor will the government need to provide supplemental funding.

In late 2011, the United States Federal Communications Commission (FCC) announced that most of the country's major cable companies partnered to join the initiative. The low prices will likely attract new subscribers who previously could not afford the cost of an Internet connection. Morgan Stanley is working with the cablecos to develop a microcredit program while partnering employment and education companies will offer specialized content to make Internet access more attractive to these users.

By the end of 2011, 41,000 families had enrolled in the program. The FCC said that it supported the partnership as a means to increase the country's broadband penetration, particularly amongst this otherwise underserved segment of the population, and praised its potential to guarantee digital literacy amongst the country's students.

In late 2013, Comcast increased the speed offered from 1.5 to 5 Mbps while introducing an online enrollment option and expanding eligibility to cover nearly 2.6 million families. At this point in time, 250,000 families across the country had already enrolled in the program, which amounted to one million individuals served.

Sources

Anderson, Nate. "Comcast's $9.99 Internet For Low-Income Families Goes Nationwide." *Wired*. Conde Nast Digital, 21 Sept. 2011. Web. http://www.wired.com/business/2011/09/comcasts-9-99-internet-for-low-income-families-goes-nationwide/.

"Cable Companies To Offer Broadband To Low-Income Households For $9.99/Month." *Deadline Hollywood*. N.p., 8 Nov. 2011. Web. http://www.deadline.com/2011/11/cable-companies-to-offer-broadband-to-low-income-households-for-9-99month/.

"Comcast Kicks Off Year Three Of Internet Essentials Broadband Adoption Program." *PR Newswire*. UBM Plc, 15 Nov. 2013. Web. http://www.prnewswire.com/news-releases/comcast-kicks-off-year-three-of-internet-essentials-broadband-adoption-program-in-virginia-connecting-more-than-17000-low-income-americans-in-virginia-to-the-power-of-the-internet-at-home-232055931.html.

Case Study 4.4: Banda Larga Popular (Brazil)

In 2009, only one-third of households within the Brazilian state of São Paulo had access to a broadband connection. Of the remaining two-thirds, nearly 60 % blamed that the high cost of Internet services prevented service acquisition. That year, the governments of the states of São Paulo, Pará, and Distrito Federal partnered together to offer low-income citizens in these districts affordable broadband. The social inclusion program, dubbed Banda Larga Popular, provided Internet connections for US$ 17 per month (35 reals, or 29 reals in those states where ICMS taxes do not apply). The Ministry of Communications oversees all monitoring and compliance while the operators are in charge of the provision of broadband and the promotion of such services. Inked in 2011, the initiative falls under the umbrella of Brazil's National Broadband Plan (PNBL) and will run through 2014.

Telefónica Brasil delivered the service, which reached speeds up to 256 kbps using wi-mesh technology. To ensure that the telco could cover its costs while charging such a low price, the government waived the 25 % ICMS tax. While the government invited other telcos to participate, only Telefónica chose to do so.

Within São Paulo, the project target low-income households that already owned a computer but did not have an Internet connection. Initial analysis estimated that this criterion would include approximately 2.5 million households.

In December 2010, Cisco announced the results of its Broadband Barometer study, showing that Brazil had reached 16.2 million fixed broadband connections earlier in the year. This number reflected an increase of 7.9 % over the previous half year and 18.1 % over 2009. The study credited the growth in home computers with the rise in broadband subscribers, which it said came as a result of lower tax rates on ICT equipment and by the increase in low-cost broadband plans available to new users. Cisco directly cited the success of the Banda Larga Popular initiative. This trend continued, and the country saw 25.8 million broadband connections by mid-2013, with this number expected to top 42.6 million by 2017.

Sources

"Brazil Achieves More Than 16 Million Broadband Connections." *The Network*. Cisco, 9 Dec. 2010. Web. http://newsroom.cisco.com/dlls/2010/prod_120910.html.

"Telefonica Brasil SA (VIV)." *Reuters*. Thomson Reuters, n.d. Web. http://www.reuters.com/finance/stocks/companyProfile?symbol=VIV.

"São Paulo Government Introduces Low-costing Broadband." *Business News Americas*. N.p., 15 Oct. 2009. Web. http://www.bnamericas.com/news/telecommunications/Sao_Paulo_government_introduces_low-costing_broadband.

"Brazil Set for More Than 42.6 Million Connections in 2017." *The Network*. Cisco, 14 May 2013. Web. http://newsroom.cisco.com/release/ 1188047/Brazil-Set-for-More-Than-42-6-Million-Connections-in-2017.

4.2.5.3 Negotiation of Low Priced Service for Public Administration Facilities (Health Care Centers, Schools, Libraries)

In this case, the offering is negotiated but only for providing broadband access at a reduced price to public entities charged with providing social services.

Bringing affordable, high-quality broadband access to rural, disadvantaged populations is crucial to any economy hoping to harness the benefits of the technology. Offering broadband in public areas like hospitals, schools, and libraries can simultaneously increase the sheer number of citizens using it while improving these institutions' quality of service. Further, in order for this population to access the educational and health care services that otherwise tend to remain concentrated in more urban locations, the most rural and remote areas of a country must have broadband connectivity. Thus, policies and programs that promote reduced-price broadband for public administration facilities can bridge the ever-widening digital divide while also allowing these facilities to have a wider reach when providing their services.

When looking at hospitals, for instance, low-quality or nonexistent broadband can "limit [their] ability to use technology to reduce costs or enhance quality of care" (Zager 2012). Quality health care depends on connectivity, but the service is not always affordable or accessible, particularly in rural areas. Lack of access puts hospitals at a disadvantage, limiting their ability, for instance, to access online resources or offer remote consultations to patients who otherwise have difficulty reaching the hospital for medical care. Further, broadband connections have the potential to attract more doctors to otherwise isolated areas, where they can use the technology to reach more patients and better serve them. With more doctors and healthcare personnel, the area then sees an increase in employment opportunities and economic activity.

Policy makers have tackled this need in various ways. The ITU advocated, for instance, for the revision of preexisting universal service programs to emphasize the provision of broadband services (GSR11 2011). Subsidies, the agency stressed, could increase access across all segments of the population while spurring innovative services without distorting the market. Financing broadband access for schools and hospitals would in turn create demand for advanced ICT connectivity and services.

Several countries have enacted policies in this domain. In the United States, the FCC provides a 65 % subsidy to rural public or nonprofit rural health care providers to use toward the cost of broadband network deployment or subscriptions through its Health Connect Fund (Mitchell 2013). In their 2008 report for the

Information Technology and Innovation Foundation, Atkinson et al. (2008) highlight many policies that support broadband supply and demand in the international community. The Canadian government's BRAND offers $80 million to eligible communities to support broadband infrastructure projects. Such projects have included Alberta's SuperNet, which links schools, libraries, healthcare facilities, and government offices, and Saskatchewan's CommunityNet, which brings broadband to public sector institutions (Atkinson et al. 2008). The Korean government spent $24 billion on a public backbone network, which service providers used to offer broadband to 30,000 government and research institutes and 10,000 schools. Likewise, Scotland's 2004 Broadband Pathfinder Project offered grants to wire schools, libraries, and public buildings.

Case Study 4.5: Projecto Banda Larga nas Escolas Publicas Urbanas (Brazil)

The government of Brazil launched its initiative *Projeto Banda Larga nas Escolas Públicas Urbanas* (PBLE) in 2008, committing to the provision of free broadband connections to a minimum of 90 % of urban schools. The communications ministry established the project as a partnership with the ministry of education and the telecom regulator, Anatel. The Brazilian telecommunications operators signed concession agreements with the government agreeing to support the program as a term of rolling out their broadband backhaul infrastructure. As part of the agreement, the operators are required to provide the broadband, an agreement that critics argue forces customers to bear the costs.

Per the communications ministry, 91.6 % of urban public schools (57,586 institutions) received free Internet connections by 2010. By October 2012, 63,394 public primary and secondary schools offered broadband access. Of these schools, 37,773 are municipal, 25,363 are state-owned, and 258 are federal. In 2012, the project connected an average of 17 schools per day. It will likely continue through 2025 to impact a total of 70,000 schools and update connection speeds, which at present are between 2 and 10 Mbps. Until the 2025 deadline, the PBLE program will continue to cover all service and maintenance fees at no cost to the schools.

Sources

"All Urban Public Schools to Have Broadband by Year-end." *Telecom. Business News Americas*, 2 Feb. 2011. Web. http://www.bnamericas.com/news/telecommunications/all-urban-public-schools-to-have-broadband-by-year-end.

"More than 63,000 Public Schools in Brazil Have Broadband." *Internet. Telecompaper*, 11 Oct. 2012. Web. http://www.telecompaper.com/news/more-than-63000-public-schools-in-brazil-have-broadband–901220.

National Broadband Plans. Rep. Organization for Economic Cooperation and Development, 11 June 2011. Web. http://www.oecd.org/internet/interneteconomy/48459395.pdf.

"Popular Broadband Has Already Reached More than 1,800 Brazilian Cities." *Porta Brasil*. N.p., 6 Sept. 2012. Web. http://www.brasil.gov.br/news/history/2012/09/06/popular-broadband-has-already-reached-more-than-1-800-brazilian-cities/newsitem_view?set_language=en

"Programa Banda Larga Nas Escolas." Ministerio Da Educação, n.d. Web. 27 Feb. 2014. http://www.educacao.gov.br/index.php?option=com_content&view=article&id=15808&Itemid=823.

4.2.5.4 Free Wi-Fi Access Points

The provision of free Wi-Fi Internet access is being conceived as one of the building blocks needed to build a city's international competitiveness. There are several features and options of a free Wi-Fi program:

- Coverage of public spaces: squares and parks, public transportation, including metros, public libraries
- Type of service: amount of time provided for free access (1 h limit while commuting, open unlimited access)
- Type of service provider: under contract with telecommunications operators or other broadband player, offered by the city administration
- Quality of service: basic 128 kbps, video-streaming quality
- Business model: free provision based on a singular event, then moving to a prepaid offering, potentially including customized interactive digital advertising

Case Study 4.6: Free Underground and Overground Internet (London)

Transport for London, which oversees the city's public transportation, has demonstrated a commitment to providing IT services. Beyond using technology to improve its operations, the agency understands that data access allows commuters to "make better travel choices." Innovations include open standards, sensor networks, and "ubiquitous connectivity."

In 2012—months before the Summer Olympic Games—Transport for London contracted Virgin Media to provide free Wi-Fi within the London Underground and the ISP Cloud for the London Overground. The project provided 200,000 daily commuters with 1 h of free Wi-Fi daily at speeds fast enough to stream video without delay.

Virgin's contract, which covered the cost of deployment, included 120 London stations, all of which had Wi-Fi access by the end of 2012. Virgin continued to offer the service for free through year-end 2012. Access to online commuting information remained free. Global firms showed interest in the project as a way to advertise their products and monitor users' browsing trends throughout London while providers could gauge the best locations for future free Wi-Fi hot spots.

In 2012 the carrier O_2 completed the deployment of Underground and Overground Wi-Fi hotspots. While users must register, they access the network free of charge. O_2 targeted local businesses for advertising revenue. As of early 2014, 130 London Underground and 56 Overground stations offer Wi-Fi. Virgin, EE, T-Mobile, Orange, O_2, and Vodafone customers can access the service for free; Virgin Media offers passes for noncustomers.

Sources

Ungerleider, Neal. "London Underground, Overground Get Free Wi-Fi." *Fast Company*. N.p., 27 Mar. 2012. Web. http://www.fastcompany.com/1826398/london-underground-overground-get-free-WiFi.

Rasmussen, Paul. "Nokia Funds Free WiFi in London for Windows Phone Marketing Campaign." *FierceWirelessEurope*. N.p., 2 Nov. 2011. Web.   http://www.fiercewireless.com/europe/story/nokia-funds-free-WiFi-london-windows-phone-marketing-campaign/2011-11-02.

"Transport for London." *CIO 100*. CIO UK Magazine, 2012. Web. http://www.cio.co.uk/cio100/transport-for-london/114294/.

Ray, Bill. "Virgin Media Snags London Underground WiFi Monopoly." *The Register*. N.p., 15 Mar. 2012. Web. http://www.theregister.co.uk/2012/03/15/virgin_wi_fi/.

"Station WiFi." *Transport for London*. N.p., n.d. Web. http://beta.tfl.gov.uk/travel-information/improvements-and-projects/station-wifi.

Case Study 4.7: Public Wi-Fi (France and Greece)

In 2006, Paris mayor Bertrand Delanoe announced that free public access Wi-Fi "is a decisive tool for international competition." 1 year later, the city launched its Wi-Fi network with access points in public parks, squares, and libraries. As part of a contract with the mayor's office, Alcatel-Lucent and wireless operator SFR partnered to build and manage the network. Paris and the Ile-de-France regions funded the project, which totaled approximately 2 million EUR plus an additional 500,000 EUR annually for maintenance. As of early 2014, the city now has 260 public Wi-Fi spots and offers a website with connection instructions for residents and visitors alike.

This initiative was not the first of its kind in Europe. With more than 2 million residents of Paris proper and more than 11 million residents in the Parisian suburbs, however, it perhaps had the largest reach. In 2005, the Greek town of Trikala launched its 80 million EUR E-Trikala initiative, which included the installation of the town's first 10 free public Wi-Fi points. In its first year alone, it attracted 3,500 users. In 2007, the second phase of the project went live, with Swedish telecom manufacturer Ericsson contracted to deploy the more expansive infrastructure and improve service. E-Trikala also partnered with Cisco to build a 15-km metro optical network linking city buildings and connecting additional hotspots. By mid-2007, the number of registered users had doubled since year-end 2005, reaching more than 6,000 users.

Following this project, seven additional Greek municipalities announced plans for similar initiatives. These cities and towns signed an agreement to connect with each other to create "the first digital community in Greece." Sweden saw a similar model, whereby more than 150 individual towns built their own networks and connected through a shared IP backbone.

Sources

Vos, Esme. "Free Wi-Fi Service in Public Areas in Paris." *MuniWireless*. N.p., 29 Sept. 2007. Web. http://www.muniwireless.com/2007/09/29/free-wi-fi-service-in-public-areas-in-paris/.

Le Maistre, Ray. "Public Wi-Fi Comes to Paris." *Light Reading*. UBM Tech, 2 Mar. 2007. Web. http://www.lightreading.com/document.asp?doc_id=118583.

"How to Access the Wi-Fi Free of Charge in Paris." *Paris*. Mairie De Paris, n.d. Web. http://www.paris.fr/english/english/how-to-access-the-wi-fi-free-of-charge-in-paris/rub_8118_actu_109289_port_19237.

Case Study 4.8: *Gov Wi-Fi* (Hong Kong)

In 2008, Hong Kong finalized its revision of the 1998 Digital 21 Strategy, stating the following objectives:

- Facilitating a digital economy;
- Promoting advanced technology and innovation;
- Developing Hong Kong as a hub for technological cooperation and trade;
- Enabling the next generation of public services; and
- Building an inclusive, knowledge-based society.

The strategy recognized the importance of investing in information infrastructure to promote economic growth. Already boasting some of the world's highest broadband penetration rates, Hong Kong looked for additional ways to encourage Internet access. Incorporated into the Digital 21 Strategy, the Government Wi-Fi Program—known as GovWi-Fi—installed Wi-Fi hotspots in government locations to offer free broadband access.

Later that year, the program began the installation process throughout all 18 districts. Locations included public libraries, public enquiry service centers, sports venues, cultural and recreational centers, cooked food markets and cooked food centers, job centers, community halls, major parks, government buildings, and offices

Per its website, through GovWi-Fi:

- People can surf the web freely for business, study, leisure or accessing government services whenever they visit the designated Government premises.
- Business organizations can extend their services to a wireless platform to reach and connect with their clients.
- ICT industry players can make use of this new wireless platform to develop and provide more Wi-Fi applications, products and supporting services to their clients, and open up more new business opportunities.
- Foreign visitors can enjoy Internet access at the designated tourist spots.

By mid-2012, the program offered services at more than 400 locations. Its website allows users to search for locations and also offers security tips and addresses the Wi-Fi-related health concerns that have been raised. To encourage use of the Wi-Fi spots, the program periodically holds events, contests, and promotion campaigns.

The Next Generation Gov*Wi-Fi* Programme commenced in 2013 following the conclusion of the initial campaign. It extended service to an additional 40 locations, increased coverage and speed, and supported the IPv6 protocol while enhancing the GovWi-Fi portal.

Sources

"Programme Overview." *GovWiFi*. GovHK, January 2014. Web. http://www.gov.hk/en/theme/wifi/program/index.htm.

"2008 Digital 21 Strategy." N.p., 22 Apr. 2008. Web. http://www.digital21.gov.hk/eng/strategy/2008/Foreword.htm

Case Study 4.9: Google's "Free Wi-Fi" Project (Brazil, India)

In December 2012, Google partnered with Enox, the Brazilian advertising firm, to launch its "Free Wi-Fi" project, allowing users to access the Internet from their personal smartphones, tablets, or computers in bars throughout seven southern cities.

Google Brazil stated that, "The number of people with smartphones in Brazil is greater than in Germany, France, and Australia, and most of them use their devices every day to read news, watch video clips and connect with their friends. By means of this project, we're sure that the Brazilians will be able to enjoy better their friends when they are at the pub, besides creating and registering memories of their moments."

The connection utilized a Wi-Fi or fiber optic connection depending on the bar, offering connections between 10 and 30 Mbps. Users did not face time limitations or browsing restrictions. Beyond offering citizens free Internet services, the project also promoted the Google brand. When users activated the Wi-Fi, their devices automatically launched the Google home page, which displayed the company's products and download suggestions. Free Wi-Fi ran for 90 days during the Brazilian summer.

The multinational Internet company had experimented with a similar project in India, partnering with O-Zone Networks to offer free Wi-Fi to increase the user base of Google+ and YouTube. While Internet use in Brazil was unlimited and unrestricted, users in India had to pay to access any sites beyond Google+ (unlimited) or YouTube (10 min per week for free). The partnership between Google and OZone also lasted 3 months.

Proponents saw its potential for increasing Internet access in a country with less than 10 % Internet penetration. Prior to the launch of this project, Google cited that 40 % of Internet searches and 67 % of e-commerce came from mobile Internet use and that free Wi-Fi access would only further encourage this trend of accessing the Internet from portable devices.

This project bears similarity to Google's collaboration with Boingo Wireless in the United States, when it sponsored free and discounted Wi-Fi throughout New York City hot spots. The offering expanded to other metropolitan areas, including Chicago, Houston, Los Angeles, New York, Seattle, and the District of Columbia.

Sources

"Google Offers Free Wi-Fi in 150 Bars in Brazil." *Telecompaper*. N.p., 14 Dec. 2012. Web. 06 Mar. 2013. http://www.telecompaper.com/news/google-offers-free-wi-fi-in-150-bars-in-brazil–914243.

Parker, Tammy. "Google Sponsoring Free Summer Wi-Fi in Brazilian Bars." *FierceBroadbandWireless*. N.p., 14 Dec. 2012. Web. 06 Mar. 2013. http://www.fiercebroadbandwireless.com/story/google-sponsoring-free-summer-wi-fi-brazilian-bars/2012-12-14.

Chan, Alice. "Google Offers Free Wi-Fi In India To Access Social Networks." *PSFK*. N.p., 11 Jan. 2012. Web. 06 Mar. 2013. http://www.psfk.com/2012/01/google-free-wi-fi-india.html.

Case Study 4.10: Free Public Wi-Fi (Thailand)

In January 2013, Thailand's regulator, the National Broadcasting and Telecommunications Commission (NBTC), announced that it had awarded a grant of US\$ 32 million from its Universal Service Obligation (USO) budget to the ICT Ministry to further its free public Wi-Fi project. The project aimed to create 40,000 public Wi-Fi spots by the year's end, with 250,000 spots slated for 2018. The initial locations included public universities and hospitals, city halls, and major tourist destinations. The installation of 150,000 access points included five access points per location with speeds of 2 Mbps per second. Each access point can accommodate 15 users at 20 min per access. By August 2013, the project had installed 270,000 free hot spots, with an additional 150,000 slated by year-end 2014.

CAT Telecom (the state-owned telecom infrastructure company) and TOT Corporation (the state-owned telecommunication operator) provide the service. The project focuses on the country's large cities and the "last-mile areas" that currently lack fiber optic networks.

The expansion of the Free Wi-Fi project will accelerate the "Smart Thailand" project, falling under "ICT2020," the government's commitment to ICT infrastructure and services. ICT2020 emphasizes universal Internet access and device affordability as well as government organization collaboration and support. Under Smart Thailand, the government committed to 80 % broadband coverage by 2016 and 95 % by 2020. In early 2012, only one-third of the population had access. Ultimately, the initiative aims to increase Thailand's global competitiveness.

The first phase of Smart Thailand runs until 2015 and will upgrade the existing telecom networks to cover the aforementioned 80 % of the population. The second phase, which runs through 2020, incorporates network installation in regions not yet covered by a fiber optic network. The National Broadband Network Company (NBN Co) will operate the nationwide network as part of a joint venture between existing private and public sector operators. The venture will reduce investment duplication and separate the installation and service provision functions. Combined, the two phases will total US\$ 2.7 billion.

Sources

Kunakornpaiboonsiri, Thanya. "Thailand to Create 300,000 More Free Wi-Fi Spots." *FutureGov Asia*. N.p., 25 Jan. 2013. Web. 06 Mar. 2013. http://www.futuregov.asia/articles/2013/jan/25/thailand-create-300000-more-free-wi-fi-spots/.

Sambandaraksa, Don. "Thailand Issues Funds for Free Public Wi-Fi." *Telecom Asia*. N.p., 17 Jan. 2013. Web. 06 Mar. 2013. http://www.telecomasia.net/content/thailand-issues-funds-free-public-wifi.

Pornwasin, Asina. "Smart Thailand Project on Track." *The Nation*. N.p., 28 Feb. 2012. Web. 06 Mar. 2013. http://www.nationmultimedia.com/technology/SMART-THAILAND-PROJECT-ON-TRACK-30176841.html.

Basu, Medha. "400,000 Free Wi-Fi Hotspots in Thailand by 2014." *Asia Pacific FutureGov*. N.p., 13 Aug. 2013. Web. http://www.futuregov.asia/articles/2013/aug/13/400000-wi-fi-hotspots-thailand-2014/.

Case Study 4.11: Public Wi-Fi (Estonia)

By 2008, nearly all of Estonia—a country that spans 45,000 square-kilometers and has a per capita GDP of just $14,300—had Wi-Fi access following a nationwide push to install access points throughout the country. Colloquially known as "E-stonia," by this point 1.4 million residents—50 % of whom resided in rural areas—were connected to wireless Internet and 70 % of the population conducted personal banking transactions online.

Beginning in 2002, volunteers with the WiFe.ee organization lobbied for local cafes, hotels, hospitals, parks, and so on to offer Internet access, working with them to design and implement the necessary networks. WiFi.ee stressed the importance of working with locals in the deployment process, as they feared outsiders would not understand the needs of the people nor the geographic or political issues that could potentially cause roadblocks. In fact, once established, nearly all wireless connections were managed by local business owners who recognized that not having Wi-Fi was akin to encouraging customers to go elsewhere.

With few exceptions, access came at no cost to the user. Further, with the exception of public schools and libraries, the entire network deployment was made possible without government assistance. Local businesses took responsibility for the creation of the more than 1100 hotspots throughout Estonia. Additionally, by 2005, nearly all schools had Internet access. From their home computers and mobile phones, students could access their schools' servers and connect to national libraries.

In 2010, Veljo Haamer, the creator of Wi-Fi.ee estimated that his organization had set up the Wi-Fi found in 75 % of the bars and cafes in Estonia's capital, Tallinn. Wi-Fi.ee charges local business owners between US$ 300 and $500 for the connection setup and maintenance. As more businesses adopted the model of offering free Internet services to customers, other businesses had to follow suit to remain competitive. As the service became more ubiquitous, even train and bus lines began offering Wi-Fi connections.

Sources

Borland, John. "Estonia Sets Shining Wi-Fi Example—CNET News." *CNET News*. CBS Interactive, 5 Nov. 2008. Web. http://news.cnet.com/Estonia-sets-shining-Wi-Fi-example/2010-7351_3-5924673.html.

Basu, Indrajit. "Estonia Becomes E-stonia." *Government Technology.* N.p., 9 Apr. 2008. Web. http://www.govtech.com/e-government/Estonia-Becomes-E-stonia.html?topic=117673.

Boyd, Clark. "Estonia's 'Johnny Appleseed' of Free Wi-Fi." *DiscoveryNews.* Discovery Communications, 11 July 2010. Web. http://news.discovery.com/tech/estonias-johnny-appleseed-of-free-wi-fi.html.

Case Study 4.12: Fon (Spain and International)

Created in Madrid in 2005, where it is headquartered, Fon Wireless is now incorporated and registered in the United Kingdom with offices in both Spain and the United Kingdom as well as in the United States, Brazil, France, Germany, and Japan. Its investors include such names as Google and Skype. Described as "crowdsourced Wi-Fi," Fon allows its members to access free roaming through Fon Wi-Fi Spots by sharing their own home Wi-Fi. In essence, these access points combine to create a "network where everyone who contributes connects for free." As of February 2014, Fon users now have access to more than 12 million hot spots.

Accessing Fon Wi-Fi requires either a Fon Wi-Fi router with a broadband connection or membership with one of Fon's telco partners, which have Fon integrated into their CPE devices and offer customers DSL/Cable modems with the preinstalled Fon feature. Nonmembers can also access a Fon Spot by purchasing an access pass once connected to the Wi-Fi signal. The passes are only available on the login page of each individual hotspot.

Each Fon Spot consists of two separate, dedicated Wi-Fi signals, one for the home user and the other for other members and visitors of the network. The home traffic is prioritized so as not to slow down home Internet use. Further, to guarantee security, a firewall separates the home signal from the guest signal. Fon also encrypts user login information.

While Fon began as a small startup in Spain, it has since grown into a service accessed around the world and now with large-scale broadband providers in several countries. In 2007, for instance, it formed a partnership with British Telecom (BT), giving the telco's three million broadband customers the option to join the network, which had nearly 200,000 global Wi-Fi hotspots—the largest network of Wi-Fi hotspots in the world at the time. Prior to BT, Fon also secured partnerships with such providers as the United States' Time Warner Cable and France's Neuf. While some Internet Service Providers continue to prevent their customers from sharing broadband, others recognize that including access to free Wi-Fi hotspots outside of the home makes broadband packages more appealing to potential customers. BT, for example, boasts that its subscribers don't just get a broadband

connection in their house, but also access "from a park bench in New York, to a bus stop in London, to an apartment in Tokyo."
Sources
"What Is Fon?" *British Telecom.* N.p., n.d. Web. http://www.btfon.com/.
"How It Works." *FON.* N.p., n.d. Web. http://corp.fon.com/how-it-works.
 Schonfeld, Erick. "Fon Inks Deal With British Telecom." *TechCrunch.* N.p., 4 Oct. 2007. Web. http://techcrunch.com/2007/10/04/fon-inks-deal-with-british-telecom/.

4.2.5.5 Service Subsidization

Under service subsidization policies, the government offers a refund on the cost of broadband access. While rare, this option for reducing broadband service cost is being increasingly examined as a complement to some income redistribution policies.

Case Study 4.13: Education Tax Refund (Australia)

The Australian Government's Education Tax Refund (ETR) offers low-income parents, caregivers, legal guardians, and independent students refunds on the education expenses of primary and secondary school children. These costs can include computers and Internet connections, as well as such items as educational software, textbooks, and school supplies. Recipients for this tax refund qualify based on their Family Tax Benefit (FTB) eligibility.

 Per the ETR website, the refund covers the expenses of home computers as well as many tools that encourage students' home computer use, such as computer equipment and repairs, home Internet connections, and textbooks.

 While the website provides an ETR calculator, it also offers guidelines in terms of what items can be claimed and the amount of the deduction. A full-time caregiver, for instance, can claim 50 % of eligible education expenses up to $794 (to receive $397) for each primary school child and $1,588 (to receive $794) for each secondary school child. The same amounts apply to individuals who share the care of a child.

 In 2011, the program cost approximately $4.4 billion and reached 1.3 million Australian families. The government used television, print, radio, and online advertising to target eligible families.

 The 2012 Budget introduced the new Schoolkids Bonus, which replaced the Education Tax Refund. Through this new program, eligible families automatically receive two payments per year of $205 for each primary

Fig. 4.14 Price and quantity indices for personal computers (1977–2004). *Source* Greenwood and Kopecky (2008)

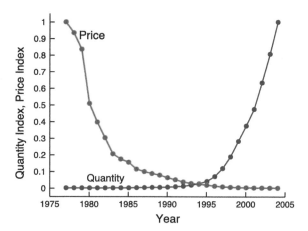

school child and $410 for each secondary school child without the need to claim expenses at tax time.

Sources

"Education Tax Refund." Australian Government, n.d. Web. 14 Dec. 2012.

Gillard, Julia. "Helping Australian Families with Back-to-school Expenses." *Australian Labor News.* N.p., 6 Jan. 2011. Web. http://www.alp. org.au/federal-government/news/helping-australian-families-with-back-to-school-ex/.

4.3 Device Ownership as a Barrier to Adoption

In the introduction of this chapter, it was indicated that broadband access requires devices capable of accessing the Internet, ranging from computers supplemented with a modem (called USB modem, dongle, or air card) to smartphones, netbooks, and tablets. Beyond service pricing, broadband economic adoption obstacles are linked to device prices. This section reviews the trends in device pricing and presents policy initiatives aimed at reducing the purchasing cost. It focuses on two areas: personal computers and mobile devices.

Before reviewing the policies that could be potentially implemented for reducing the access cost of devices, it is important to mention that device pricing has been consistently declining driven by a reduction of production costs (manufacturing economies of scale and component costs) and increasing demand. The decline has been even more abrupt if quality and performance improvements are

Table 4.14 Price performance of the Apple iPhone (by model)

Date	Model	Price	Memory (GB)	Features
June 2007	iPhone	$499, $599 (only through contract with ATT)	4, 8	Multitouch screen, up to 16 GB of storage, 620 MHz processor, 2-megapixel camera for still images, USB dock
July 2008	iPhone 3G ("iPhone 2")	$199, $299 ($599, $699 without contract)	8, 16	Assisted GPS;
June 2009	iPhone 3GS	$199, $299 ($599, $699 without contract)	16, 32	Up to 32 GB of storage, 833 MHz processor, 3.0-megapixal camera that included video recording, digital compass, voice control, Nike+
June 2010	iPhone 4	$199, $299 ($599, $699 without contract)	16, 32	Bigger battery, 1 GHz processor, 5.0-megapixel camera with LED flash, HD video recording
October 2011	iPhone 4S	$199, $299, $399	16, 32, 64	Intelligent voice recognition; an A5 processor; an 8-megapixel camera; dual antennas so that it can be used almost anywhere in the world
September 2012	iPhone 5	$199, $299, $399	16, 32, 64	20 % lighter than the iPhone 4S; an A6 processor, a new dock connector, LTE data connection, a boosted audio system, 8-megapixel Face Time HD; longer battery life, improved graphics.

Source http://abcnews.go.com/Technology/iphone-features-apple-smartphone-evolved-introduction-2007/story?id=17226964#3
The 16 GB is now available for $99 with a new carrier contract

factored in. For example, Greenwood and Kopecky (2008) compared the personal computer price index with its quality index between 1977 and 2004, and pointed to a 25 % per year decline in quality-adjusted prices with an equal rise in demand (see Fig. 4.14).

Retail pricing of personal computers ranges currently between US$ 275 and US$ 1,600, a decline from US$ 800–US$ 2,500.[15] Moreover, prices for refurbished devices can reach between US$50 and US$ 100.

The decline in smartphone prices has also been significant, if not more dramatic than in the case of personal computers. Table 4.14 presents pricing and performance data for the different generations of the popular Apple iPhone.

While the introductory subsidized price of the first iPhone model was US$ 499 (with 4 GB of memory), the current model (iPhone 5) starts at US$ 199 (with a minimum of 16 GB and a wide range of features).

Even if the iPhone is targeted at higher income segments in the emerging world, manufacturers such as Huawei and Nokia are offering low-priced smartphones. For example, Huawei in partnership with Safaricom launched an US$ 80 Android phone in Kenya. Still, carriers in the emerging world consider that $50 is a suitable price point for a smartphone.

In late-2012, the Samsung Galaxy overtook the iPhone as the world's best selling smartphone, a reflection of its efforts to target a larger user base though phones with a broad range of features at various price points. Apple has released six models of the iPhone since 2007, with each version seen as a successor to the previous model. In contrast, since 2009 Samsung has released more than 20 versions of the Galaxy with models and prices varying for targeted demographics. Much of the success of the phone has been credited with its ability to simultaneously market the high-end models in developed economies while pushing cheaper models to capture key markets like India, China, and Brazil.[16]

The iPhone operates on the iPhone Operating System (iOS), but all Galaxy phones feature the Android operating system and thus compete with other Android smartphones. Considering that certain Internet access devices still remain out of reach of disadvantaged segments of the population in the emerging world, governments have at their disposal a range of policy models aimed at tackling the device affordability barrier. The following sections review first programs aimed at reducing the purchasing cost of computers, and then present initiatives focused on handsets, primarily smartphones.

---

[15] The introductory price of an Apple Macintosh 128 K was US$2,495 (8 MHz CPU speed, 128 K DRAM, 64 K Rom).

[16] http://readwrite.com/2012/04/26/a-brief-history-of-the-samsung-galaxy?&_suid=1363009792 9870603560470510 2742 and http://www.intomobile.com/2012/02/17/npd-lowcost-android-smart phones-seize-80-emerging-markets/.

4.3.1 Programs to Reduce the Cost of Purchasing Personal Computers, Laptops, and Netbooks

Three types of programs have been implemented to overcome the personal computer ownership barrier. The first one focuses on the provision of subsidies to reduce the acquisition price of devices. The target in this case could be households at the lower end of the sociodemographic pyramid, primary school to university students, and SMEs (especially micro-enterprises).

The second program typically targets students in primary education, with governments distributing "One Computer per Child." In this case, public school students receive computers free of charge.

The third type of initiative entails a reduction of the access price by eliminating or decreasing taxes paid at time of purchasing. Levies affected by this measure could range from sales tax, import duties, and even sector-specific levies.

4.3.1.1 Targeted Subsidies

While tax reduction could be an indirect subsidy, this section will address initiatives such as vouchers or the provision of lower priced devices for qualifying segments of the population (e.g., students).

Case Study 4.14: Computer Subsidies (China)

In 2009, more than half of China's population lived in the rural parts of the country, where the average per capita annual income was US$ 700 (25 % the average income of urban residents), broadband penetration rates were lower than in the urban areas, and the personal computer market was nearly untapped. Further, PC shipments decreased globally and spending fell.

To stimulate rural spending, the Chinese government launched a subsidy program offering a 13 % rebate to rural residents buying select products to help PC manufacturers increase their sales to the country's under-developed regions, particularly after national computer demand fell. The rural computer subsidy came as part of a larger US$ 586 billion subsidy program to increase demand for home electronics, known as the Home Appliance Subsidy Program.

The government identified 14 vendors that could participate in the program and sell low-priced PCs in rural China, making computers more affordable while also spurring industry competition. These manufacturers created special products for the program with two-thirds of the computer models priced under US$ 500. The products also met regional specific demands. The PCs, for instance, kept potential variations in power supply voltage—a frequent problem in rural areas—in mind. Further, many of them came with special software for farmers, like inventory management programs. Vendors also ensured physical proximity to their customers, as citizens in rural areas did not have the means or the desire to drive for hours to buy a computer.

Hewlett-Packard sponsored variety shows and film screenings and offered product demonstrations in small towns. It also sent buses equipped with its products to elementary schools to advertise and to train students on how to use the technology. Competitor Lenovo began marketing its computers as luxury wedding gifts, employing the slogan, "Buy a Lenovo PC, Be a Happy Bride," and delivering them in large, conspicuous boxes. The company also has a flashy showroom with a section of the store devoted to products designed specifically for rural use.

Nearly 60 % of all rural residents—or 200 million households—qualified for a subsidy. Initial estimates expected the program to generate the sales of 800,000 computers. The program ended in early 2012 and was dubbed a success. Combined, the subsidies covering all electronic goods for farmers and rural residents generated a 53 % increase in sales.

Sources

Chao, Loretta. "PC Makers Cultivate Buyers in Rural China." *Tech Journal*. Wall Street Journal, 24 Sept. 2009. Web. http://online.wsj.com/article/SB125366214543432237.html.

Lemon, Sumner, and Owen Fletcher. "China Offers Computer Subsidy for Farmers." *Desktops*. PCWorld, 5 Mar. 2009. Web. http://www.pcworld.com/article/160750/article.html.

He, Helen, and Simon Ye. "Rural China PC Program Will Increase PC Shipments in 2009 | 909330." Gartner, 10 Mar. 2009. Web. http://www.gartner.com/id=909330.

"China Launched A Massive Subsidy Program To Get People To Buy Appliances." *Business Insider*. N.p., 18 Jan. 2012. Web. http://www.business insider.com/chinas-successful-appliance-subsidies-at-an-end-2012-1.

Case Study 4.15: Kenniswijk Project (Netherlands)

The Dutch General Directorate of Telecommunication and Post (DGTP) partnered with the Ministry of Economics to develop the Kenniswijk Project in 2001. The project provided residents of Eindhoven, Helmond, and Nuenen with computers, mobile, and Internet products 2 years ahead of the rest of the country free of charge, creating a "consumer market of the future."

Deemed "a vision of the broadband future," this experiment allowed the ministry to examine the social and economic effects of increased ICT and broadband access within a closely monitored community. Analysis of the project served as a best practices guide and shaped the development of future technology while promoting a synergy between infrastructure deployment and service provision.

The project ultimately grew into collaboration between 27 private and public parties. The organization oversaw the facilitation, motivation, and support of the involved organizations. Participants received TV/video, telephony, and broadband services. Multiple services providers could access the physical network, stimulating competition, and innovation, particularly in areas such as e-health and distance learning.

The government allocated US$ 40.5 mn toward the project. US$ 11 mn went toward infrastructure development subsidies, which included a US$ 700 discount per user and more than 15,000 connections. Participants in the program received a free 10 Mbps Internet subscription for the first year and were not charged additional connection costs. The 10,000 subscribers received regular newsletters and had access to the project's website and helpdesk. Volunteer organizations and visitor centers were also encouraged to take advantage of subsidies to put the technology to use.

In total, the endeavor spurred approximately 1000 new project ideas, 300 of which were turned into concrete project plans and subsidy requests. 135 were ultimately approved.

The project ended in October 2005, resulting in 15,000 total FTTH connections and 135 services in Kenniswijk. Through this experiment, policy makers learned a great deal regarding motivating companies to install infrastructure and users to subscribe to a service. Even once the project ended, more than 80 % of households continued to use the fiber connection. Following the success of this project, nearby towns implemented similar programs based on the same approach.

Sources

"Dutch National Project for Broadband Innovation Selects PacketFront for Its next Generation FTTH Platform." PacketFront, 10 Nov. 2004. Web. http://www.packetfront.com/en/news_events/press_releases/2004/009.html.

Kramer, René, Alex Lopez, and Ton Koonen. "Municipal Broadband Access Networks in the Netherlands." Proc. of AccessNets, Athens. Breath, 4 Sept. 2006. Web. http://w3.ele.tue.nl/fileadmin/ele/TTE/ECO/Files/Pubs_2006/Kramer_AccessNets_06_presentation.pdf.

4.3.1.2 Distribution of Free Devices

These programs are more prevalent with regards to computer distribution, although they could be extended to other broadband access devices such as smartphones.

Case Study 4.16: SchoolNet Project (Namibia)

Established in 2000 to offer sustainable, low-cost ICT and Internet services to all Namibian schools, the not-for-profit SchoolNet Namibia also provided training to empower youth through ICT. In partnership with local telcos and international development agencies, it brought affordable computers and solar-power labs to schools while promoting open source software, Creative Commons educational content, and discounted flat-rate wireless Internet services. Open source software offered cost-reduction and scalability advantages and made sharing and adaptation easier for local communities. SchoolNet also created its own ISP, XNet, for schools to access the Internet through dial-up or wireless connections.

The national government recognized SchoolNet's contribution to ICT rollout and job creation, in line with its National Development Plans. The organization served as an example for subsequent sustainable ICT education projects across Africa.

Multiple international aid agencies supported the project, with The Swedish International Development Cooperation Agency offering assistance totaling US\$ 3 mn. Telecom Namibia offered schools a flat-rate Internet package for US\$ 25 per month. Schools that could not afford services qualified for cross-subsidies from schools with more resources. Volunteers provided basic ICT support. Each lab held five refurbished computers plus a server and a printer.

In 5 years, the project set up computer labs and connections to the SchoolNet ISP in 300 schools while connecting libraries, teacher resource centers, and NGOs. By the project's 2009 end, the ISP boasted more than 180,000 regular Internet users, compared to the 11,000 users of the next largest commercial provider.

Sources

"SchoolNet Namibia." *Panafrican Research Agenda on the Pedagogical Integration of ICTs*. ERNWACA, 28 Feb. 2009. Web. http://www.ernwaca. org/panaf/spip.php?article567.

   "Broadband Strategies Handbook." Ed. Tim Kelly and Carlo M. Rossotto. The World Bank, 2012. Web. https://openknowledge.worldbank.org/handle/10986/6009.

   "The Case of SchoolNet Namibia." *WikiEducator*. N.p., 9 Oct. 2010. Web. http://wikieducator.org/The_Case_of_SchoolNet_Namibia.

   Du Buisson, Uys, and Chris Morris. "Schoolnet Namibia." *First Mile First Inch*. N.p., Mar. 2005. Web. http://www.fmfi.org.za/wiki/index.php/Schoolnet_Namibia.

Case Study 4.17: Federal Computers for Learning Program (United States)

In 1996, President Bill Clinton signed Executive Order 12999 "to ensure that American children have the skills they need to succeed in the information-intensive twenty first century." It led to the enactment of the Federal Computers for Learning Program, which donated "retired" federal government computers and ICT equipment to eligible schools and education-related nonprofit organizations free of charge. The program targeted prekindergarten (age 4) through high school (age 18) students within predetermined "federal rural empowerment zones" marked by high poverty and unemployment rates. Thus, the program promoted sustainability and development in disadvantaged communities.

Many of the schools face shrinking budgets and cannot afford modern computers. At the same time, the federal government disposes of approximately 10,000 computers—many of which are only 3 years old—every week. By using this equipment, the program also supports current President Barack Obama's commitment to "zero waste in government."

When the program first began, most schools received IBM-compatible PCs, though Pentium-based systems and occasional Apple computers were donated as well. Beyond computers, the government agencies also donated modems, routers, services, ICT equipment, and research technology. The schools and organizations only paid for the shipping and handling costs and, when applicable, the refurbishing costs.

The United States General Services Administration (GSA) facilitates and sponsors the program, but schools must apply for the program themselves through the Computers for Learning website. Once registered, the website enables schools to create ICT-related plans, request new computers, and find assistance, amongst other tools.

The CLP website offers a "Success Stories" section, where schools can write into share the impact the computers have had. One 9th grade teacher from Maryland wrote that after spending just 5 min registering on the site, the school received 252 computers, equipping every classroom with a computer and allowing teachers to incorporate the technology into their lesson plans. Other teachers commented on the quality of the computers, noting that not one required repair or lacked sufficient speed or memory capacity.

By mid-2012, the program donated nearly 360,000 computers and equipment—worth more than US$ 317 million—to thousands of schools and organizations throughout the country.

Sources

Longley, Robert. "Surplus Computers: Free for Schools." *US Government Info*. About.com, n.d. Web. http://usgovinfo.about.com/library/weekly/aa060901a.htm?p=1.

Computers For Learning. U.S. General Services Administration, n.d. Web. http://computersforlearning.gov/.

"Computers for Learning Puts Information Technology in Classrooms."
U.S. General Services Administration, 12 June 2012. Web. http://www.gsa.
gov/portal/content/136867.

Case Study 4.18: Smartphones to the Homeless (Republic of Korea)

In 2011, the Seoul government launched its "Smart Seoul 2015" program,
expressing its commitment to "making Seoul the 'best smart technology' city
in the world by 2015." The program addresses infrastructure expansion and
information security as well as ICT services and e-government programs. To
promote e-government services, for example, it focuses on communication
with citizens, convenience, and a campaign against the negative aspects of
increased information access. At the time, Seoul's e-government had already
ranked first place amongst the World's 100 Cities for four consecutive years,
serving as "a benchmark for countries and cities around the world."

With an emphasis on ICT and smartphone use amongst all segments of
the population—regardless of factors such as age or income—Smart Seoul
2015 incorporated programs to distribute technology to disadvantaged
groups. For instance, the city provides Braille terminals and magnifying
devices to the disabled and the visually impaired. The program also offers
PC repair, Internet addiction projects, and training courses.

In August 2012, the government announced that it would provide
homeless shelters with free smartphones and wireless Internet services. The
provision of ICT fell under the government's larger Social Networking
Service Education Program, which focused on training and skill building
amongst the city's homeless population. The smartphones allowed these
citizens to search for employment opportunities from any location, as well as
reconnect with their families and friends and interact with other members of
the community.

Citizens of Seoul could donate their used smartphones to the Seoul
Metropolitan Government, which then distributed the devices to the residents of the shelters. The government also offered workshops at the shelters,
which residents completed prior to receiving the phone. The training
encouraged utilization of social networking services and mobile applications. Each smartphone came with a US$ 20 credit, but the phones will still
work in free Wi-Fi areas once all credit is used.

Sources

Africa, Clarice. "Seoul to Provide Homeless Residents with Smartphones
and WiFi." FutureGov, 31 Aug. 2012. Web. http://www.futuregov.asia/
articles/2012/aug/31/seoul-provide-homeless-residents-smartphones-and-w/.

Sung-Mi, Kim. "Seoul Proves Value of Advanced E-Government." Korea IT Times, 9 Jan. 2012. Web. http://www.koreaittimes.com/story/19299/seoul-proves-value-advanced-e-government.

Case Study 4.19: Reaching the Third Billion—Bringing the Prepaid Miracle to Broadband

The Intel World Ahead Program, inspired to enable more first-time users online, and by the low cost prepaid mobile broadband programs in Sri Lanka and Vietnam, recognized the opportunity for an affordable solution that includes connectivity, a computing device, and beneficial content, to address all of the demand side gaps that consumers face. In fact, through initial research, it was discovered that in the total cost of ownership, the largest affordability gap was actually the cost of the broadband subscription, when compared over a 4-year period. In fast-growing developing countries (such as Brazil, China, Indonesia, Malaysia, Mexico and Russia), broadband access can account for 60–80 % of the total cost of ownership of a PC. Often, only about 20 % of citizens could afford the monthly plans. Piloting the program in early 2011, working with service providers, equipment providers, content providers, and governments, new low cost solutions were created, including entry-level notebooks, compelling content, and prepaid broadband, accompanied by exciting advertising, branding, and marketing, to enable and encourage first-time users to get online.

The pilot results were impressive, and by the end of 2011, all eight pilots were complete, enabling more than one million people with entry-level PCs, content, plus prepaid broadband packages. These programs also encouraged the PC industry to aggressively lower prices to as low as $200, and encouraged content providers to create exciting new content. In one example in Vietnam, the telecom companies, Viettel and VNPT, offered 700 MB of data download for just $2 prepaid. At that price, broadband affordability surged from 12 to 70 % of citizens. Early results from the program also showed that stimulating demand is more than just price; it's also about delivering meaningful content and applications. For example in Kenya, Safaricom package include not only entry-level netbooks, and prepaid broadband but valuable content, including British Council 'Learn English' software, Education applications such as Intel® Skoool and Encyclopedia Britannica, as well as McAfee safety applications. In addition, they come with 1.5 GB of free data download. This is a very compelling offer that enhances education and learning, and runs far better on a PC than over a phone or tablet.

Today, successful programs are running in over 45 countries worldwide, and have enabled over 15 million new users to get online. With the Total Cost of Ownership (TCO) often reduced to 2/3 of the previous cost, over one billion people can now afford to enjoy technology benefits for the first time.
Sources
The state of Broadband 2012: Achieving Digital Inclusion For All, Broadband Commission Report, September, 2012 http://www.broadband commission.org/Documents/bb-annualreport2012.pdf.

"Intel World Ahead Program: Connectivity." *Intel*. N.p., n.d. Web. http://www.intel.com/content/www/us/en/world-ahead/intel-world-ahead-program-connectivity.html.

4.3.2 Programs to Reduce the Cost of Purchasing Mobile Devices, Smartphones, and Tablets

This section addresses the issues that are similar to those of computer devices, but for mobile access equipment.

Case Study 4.20: Free Devices (South Africa)

South Africa's three main providers offer contract subscribers free or heavily discounted devices to stay competitive. Initially, this subsidy attracted new customers, but with mobile phone penetration rates exceeding 100 %, providers are reconsidering this strategy. The subsidies also make market entry difficult. When Cell C entered the market in 2001, it had no choice but to offer subsidies, negatively impacting early earnings. Once operators recoup these costs, the contract expires and customers expect a new phone.

Further, the subsidies are used to attract high-end subscribers as ARPU decreases with increased penetration amongst all socioeconomic classes. Lower end, prepay customers can also acquire phones through the second hand market.

As providers expand their operations into high-risk regions in Africa, they accrue large debts and cannot afford to lose money in more lucrative South Africa. While device subsidies may have initially increased mobile phone penetration in South Africa, they may now hold back market growth in neighboring regions. As governments impose taxes and tariffs, operators must cut costs and profitability trumps subscriber numbers.

In the meantime, however, subsidized phones may soon lead to higher profit margins in South Africa as customers utilize high-value services like mobile broadband and apps on smartphones, which most consumers cannot afford without a discount. In many African countries, disposable income serves as the largest barrier to broadband uptake. Smartphone subsidies could in turn increase Internet and broadband access in this region. If these subsidies prove too costly to justify, providers could experiment with different pricing packages or partner with the government to make the devices more affordable.

With providers investing heavily in their high-speed networks to increase capacity and encourage the demand for high-speed broadband, declining prices and incentives like device subsidies will likely contribute to a growing broadband market.

Sources

"SA's Mobile Companies May Reconsider Cellphone Subsidies." *Issue No 174*. Balancing Act, 12 Jan. 2012. Web. http://www.balancingact-africa.com/news/en/issue-no-174/telecoms/sa-s-mobile-companie/en.

Mobile in South Africa—Contributing to Improved Economic and Social Outcomes. Rep. Indepen (for Vodacom), 8 Sept. 2005. Web. http://www.indepen.uk.com/docs/mobile-in-south-africa.pdf.

Mochiko, Thabiso. "African Numbers Driven by Rise in Mobile Voice, Internet Services." *Business Day Live*. N.p., 10 Oct. 2012. Web. http://www.bdlive.co.za/business/technology/2012/10/10/african-numbers-driven-by-rise-in-mobile-voice-internet-services.

Case Study 4.21: Device Subsidies (United States)

To make broadband contract plans more appealing, many telcos also subsidize modems, computers, and tablets with high-data contract agreements. Subscribers may qualify for devices with no upfront charge in exchange for committing to long-term plans that promote the use of high-value services like Internet access. The revenue generated by these services justifies the providers' cost of the equipment, and contracts usually have high early termination fees to guarantee that they can recoup the cost.

In 2008, for instance, AT&T Mobility began subsidizing laptops. As a result, data revenue in the fourth quarter of that year rose 51.2 % compared to 4Q2007.. To do so, AT&T partnered with Lenovo to offer a US\$ 150 discount on laptops with embedded 3G modems for with a 2-year mobile data plan. To qualify for this offer, users signed up for AT&T's \$60/month service, "DataConnect." The embedded modems could handle download speeds up to 1.7 Mbps and uplink speeds up to 1.2 Mbps, but only worked with DataConnect. As part of the contract, users could terminate the agreement early for \$175.

By 2012, however, this practice no longer produced the same returns due to market saturation. While smartphone subsidization continues to make sense, the practice does not hold for the tablet market, where consumers tend to buy models designed for Wi-Fi and not cellular use. In August, AT&T announced that it would no sell tablets at subsidized prices. Tablet data plans no longer have a built-in commitment and are slightly cheaper than previous offers, with the price increasing based on data; 3 GB plans cost $30 per month while 5 GB plans run for $50. Shared data plans run for $10/month plan.

In July 2012, competitor Verizon announced that it would no longer offer unlimited data plans to new customers. While the operator insisted that subscribers would benefit from shared data plans, amidst criticism, it reversed this decision. To compensate, subscribers who receive subsidized devices will no longer have the option for an unlimited plan. In contrast, smaller providers Sprint and T-Mobile continue to offer unlimited data plans, with Sprint launching an "unlimited for life" plan in 2013.

Sources

"Broadband Strategies Handbook." Ed. Tim Kelly and Carlo M. Rossotto. The World Bank, 2012. Web. https://openknowledge.worldbank.org/handle/10986/6009.

Chacos, Brad. "AT&T Stops Offering Subsidized Tablets." *LAPTOP*. N.p., 20 Aug. 2012. Web. http://blog.laptopmag.com/att-stops-offering-subsidized-tablets.

Reardon, Marguerite. "Verizon: You Can Have Unlimited Data… Just No Device Subsidies." *CNET News*. CBS Interactive, 17 May 2012. Web. http://news.cnet.com/8301-1035_3-57436642-94/verizon-you-can-have-unlimited-data-just-no-device-subsidies/.

Perez, Marin. "AT&T, Lenovo Offer Subsidized Laptops." *Information week*. N.p., 30 Oct. 2008. Web. http://www.informationweek.com/mobility/3g/att-lenovo-offer-subsidized-laptops/211800321.

Bennett, Brian. "Sprint Officially Outs New Unlimited Plans." *CNET*. N.p., 13 Jan. 2013. Web. http://reviews.cnet.com/8301-6452_7-57593355/sprint-officially-outs-new-unlimited-plans/.

Case Study 4.22: Operator Partnerships (Uganda)

In an effort to increase their subscriber base, Ugandan service providers and operators entered into partnerships to offer lower priced devices and services. Making such tools more affordable in turn makes broadband less cost-prohibitive, in turn boosting the country's overall uptake of the technology.

In 2012, international semiconductor manufacturer Intel began working with developing market telcos to offer high-speed broadband and affordable computers. This strategy should reduce the Internet divide and strengthen these countries' economies. The model replicates the prepaid model that has boosted the demand for mobile phones, attracting first time buyers who can only pay for services they can afford at the time.

As part of this strategy, Intel partnered with Orange Uganda in May 2012 to offer subscribers a variety of lower priced personal computers and high-speed broadband packages—all available at Orange retail stores. Intel will experience growth opportunities in the computing sector while Orange will more easily reach new users and increase subscriptions. As a result of increased access to ICT and consumer spending, Uganda will likely see a rise in GDP and digital literacy as well as integration of ICT into schools. Further, as ICT access and skills increase, the country will become more competitive in the global knowledge economy.

This model is not the first time Uganda has seen partnerships between service providers that led to increased broadband affordability and adoption. In 2006, four of the country's Internet Service Providers (ISPs)—Bushnet, SpaceNet, Africaonline, and One2Net, all members of the Uganda ISP association—worked together to consolidate their purchase of bandwidth. This consolidation led to a 25 % reduction in cost, which enabled them to reduce the rates of bandwidth purchased from them by schools by 75 %.

Prior to this partnership, each ISP had acquired approximately 5–6 Mbps per month for $5,000–$6,000. When working together, they together acquired 50 Mbps at a wholesale price for buying in bulk. The providers could then pass these savings along to potential customers. One2Net, for example, had previously charged a flat rate of $600 per month for 64 K speeds. Following the consolidated purchase, it reduced the price to $150 for individuals and $250 for businesses and offered higher speeds.

At the time, the price of Internet to average individual users was prohibitively high and blamed for the country's low broadband penetration rates. Users faced charges of approximately 14 cents per minute as a result of expensive bandwidth, which the ISPs also said hurt their business.

Sources

"Intel Partners with Orange Uganda to Increase Access to "broadband PCs"" *CIO East Africa*. N.p., 9 May 2012. Web. http://www.cio.co.ke/news/top-stories/intel-partners-with-orange-uganda-to-increase-access-to-%22broadband-pcs%22.

"Uganda ISPs Join Forces to Purchase Cheaper Bandwidth." *Issue 315*. Balancing Act, 2006. Web. http://www.balancingact-africa.com/news/en/issue-no-315/internet/uganda-isps-join-for/en.

Table 4.15 Service taxation approaches by country

Continent	Universalization of service	Direct taxation without sector discrimination	Direct taxation and sector specific taxes	Service tax revenue maximization
Africa	Angola Botswana, Lesotho, S. Leone, Swaziland.	Cameroon, Chad, Cote d'Ivoire, DR Congo, Egypt, Ethiopia, Gabon, Gambia, Guinea, Guinea-Bissau, Malawi, Mauritania, Mauritius, Morocco, Mozambique, Rwanda, Seychelles, S. Africa, Zimbabwe	Burkina Faso, Ghana, Nigeria, Rep. Congo, Tunisia	Kenya, Madagascar, Senegal, Tanzania, Uganda, Zambia
Middle East	Syria, Yemen		Iran, Jordan	Turkey
Asia	Bangladesh, Nepal, Pakistan	Bhutan, China, Indonesia, Lao, Malaysia, Myanmar, P. N. Guinea, Thailand, Vietnam	India, Philippines, Samoa	Cambodia, Sri Lanka
Pacific				
Latin America	Dominican Republic, Ecuador, Venezuela	Paraguay	Bolivia, Chile, Guatemala, Nicaragua, Peru, Trinidad and Tobago	Argentina, Brazil, Colombia, Mexico
Eastern Europe	Ukraine		Azerbaijan, Georgia, Kazakhstan, Russia, Uzbekistan	
Western Europe	Albania, Greece	Austria, Bulgaria, Cyprus, Czech Rep., Denmark, Estonia, France, Finland, Germany, Hungary, Ireland, Italy, Latvia, Lithuania, Luxembourg, Malta, Netherlands, Poland, Portugal, Romania, Slovakia, Slovenia, Spain, Sweden, UK		

Source Katz et al. (2011)

4.4 Taxation as a Barrier to Adoption

The total cost of ownership of broadband is impacted by numerous taxes. On the services side, three exist:

- *Value-added tax*: most countries impose some form of value-added tax, a general sales tax or similar consumption tax as a percent of the total bill
- *Telecom specific taxes*: some countries charge an additional special communications tax as a percent of the service bill
- *Fixed taxes*: in addition to the tax as a percentage of usage, some countries charge a fixed tax that could be either driven by general communications usage or wireless usage

In addition to service-based taxes, other levies can be imposed on access equipment:

- *Value-added tax*: these represent the taxes paid directly by the consumer at time of purchasing a personal computer or purchasing and/or exchanging a smartphone
- *Customs duty*: this tax on imported equipment is already included in the retail price of the computer or the smartphone
- *Other taxes*: telecommunications specific taxes on smartphones or computers (e.g., royalties calculated on the cost of the equipment)
- *Fixed taxes*: special fixed duties on smartphones, such as ownership fees

Countries do not follow a uniform approach to mobile services taxation. While all countries tax both services and equipment, the type of taxes selected and their amount vary significantly, with the consequential varying impact on total cost of ownership of a broadband device.

A scan of service taxation approaches across countries yields four categories:

- *Universalization of service*: Reduce taxes as much as possible to stimulate broadband adoption
- *Direct taxation without sector discrimination*: high VAT while recognizing the distortion effect of sector-specific taxes
- *Direct taxation and sector specific taxes*: combine VAT with a sector specific levy
- *Service tax revenue maximization*: leverage broadband communications as a source of direct taxation, by combining high VAT, high sector specific taxes and/or a fixed levy.

While most developed and some developing nations reduce service taxes to promote universalization of broadband service, the pattern is not consistent across emerging countries. For example, the Africa and Asia Pacific continents comprise numerous nations with taxation approaches aimed at universalizing mobile services, while this approach is significantly less prevalent in Latin America (see Table 4.15).

Table 4.16 Handset taxation approaches by country

Continent	Sector discrimination based on moderate import duty	Sector discrimination based on high import duty	Sector discrimination based on high VAT and import duty but low handset specific tax	Handset revenue maximization
Africa	Angola, Egypt, Ethiopia, Gabon, Guinea–Bissau, Kenya, Mauritania, Mauritius, Morocco, Seychelles, S. Leone, S. Africa, Swaziland, Tanzania, Uganda, Zambia, Zimbabwe	Cameroon, Chad, DR Congo, Gambia, Guinea, Malawi, rep. Congo, Rwanda	Botswana, Burkina Fasso, Cote d'Ivoire, Madagascar, Mozambique, Senegal, Tunisia	Ghana, Nigeria, Lesotho
Middle East	Jordan		Turkey, Yemen	Syria
Asia	Bangladesh	Cambodia, Lao, Malaysia, Myanmar, P. N. Guinea, Pakistan, Philippines, Thailand, Vietnam	Bhutan, China, Indonesia, Samoa, Sri Lanka	India, Nepal
Latin America		Bolivia, Chile, Colombia, D. Republic, Ecuador, Guatemala, Nicaragua, Paraguay, Peru, México	Argentina, Trinidad and Tobago, Venezuela	
Brazil				
Eastern Europe		Albania, Kazakhstan, Russia, Ukraine, Uzbekistan	Azerbaijan, Georgia	
Western Europe		Austria, Bulgaria, Cyprus, Czech Rep., Denmark, Estonia, France, Finland, Germany, Greece, Hungary, Ireland, Italy, Latvia, Lithuania, Luxembourg, Malta, Netherlands, Poland, Portugal, Romania, Slovakia, Slovenia, Spain, Sweden, UK		

Source Katz et al. (2011)

Moving now to equipment (personal computers, smartphones) taxation approaches, four types can be identified, partly driven by the existence or not of import duty:

- *Sector discrimination based on moderate import duty*: VAT combined with low import duty
- *Sector discrimination based on high import duty but no telecom tax*: high import duty and VAT but no sector specific taxes on handsets
- *Sector discrimination based on high VAT and import duty but low handset specific tax*: combine high VAT with a sector specific levy
- *Equipment tax revenue maximization*: leverage broadband communications as a source of direct taxation, by combining high VAT, high customs duty and a high sector specific levy, or low import duty and high sector specific tax.

The most prevalent equipment taxation model around the world is based on VAT and, in some cases, low sector discrimination through moderate import duty (see Table 4.16).

The combination of service and equipment taxation approaches yields four distinct models:

- *Universalization and protectionism*: this approach aims at reducing levies with the purpose of decreasing total cost of ownership and stimulating broadband adoption; it can include an equipment import duty and a sector specific tax (which is relatively low and therefore has minimum distortion potential)
- *Protectionism*: this approach is similar to the one above, except that high value-added taxes on service increase substantially the total cost of ownership
- *Sector distortion*: this approach introduces sector specific service taxes with the objective of increasing government revenues but, in doing so, plays an economically distortion role by emphasizing taxes on the telecommunications sector
- *Tax maximization and sector distortion*: sector specific taxes are introduced not only on broadband services but also on equipment with the purpose of maximizing government revenues, with the consequent distortion impact.

As pointed out before, prevalent taxation models tend to differ by region. As expected, most developed countries have adopted universalization and protectionism tax approaches given that they do not need to rely on the telecommunications industry to increase revenues for the treasury. In addition, there are a number of emerging countries, which have chosen a Universalization and Protectionism approach in order to stimulate telecommunications service adoption. Notable examples in this category are China, Angola, and Malaysia.

In the next category of taxation approach—protectionism—several emerging countries that have adopted proactive ICT development strategies: India, Rwanda, Egypt, Chile, and Kazakhstan can be identified. In other words, the first two taxation categories are associated with technology development objectives.

At the other end of the spectrum there are also some significantly large emerging countries—Mexico, Argentina, Brazil, Venezuela, Nigeria, Bangladesh,

Table 4.17 Overarching taxation approach by country

	Universalization and protectionism	Protectionism	Sector distortion	Tax maximization and sector distortion
Africa	Angola, Botswana, Lesotho, S. Leone, Swaziland	Cameroon, Chad, Cote d'Ivoire, DR Congo, Egypt, Ethiopia, Gabon, Gambia, Guinea, Guinea–Bissau, Madagascar, Mauritania, Mauritius, Morocco, Mozambique, Rwanda, Seychelles, S. Africa, Zimbabwe	Kenya, Tanzania, Uganda, Zambia	Burkina Faso, Ghana, Madagascar, Nigeria, Senegal, Tunisia
Middle East	Syria, Yemen		Jordan	Iran, Turkey
Asia Pacific	Bhutan, China, Indonesia, Lao, Malaysia, Myanmar, P. New Guinea, Thailand, Vietnam	India, Philippines, Samoa	Cambodia	Bangladesh, Nepal, Pakistan, Sri Lanka
Latin America	Paraguay	Bolivia, Chile, Colombia, Guatemala, Nicaragua, Peru, Trinidad and Tobago	Dominican Rep., Ecuador, Mexico	Argentina, Brazil, Venezuela
Eastern Europe		Azerbaijan, Georgia, Kazakhstan, Russia, Uzbekistan	Albania, Ukraine	
Western Europe	Austria, Bulgaria, Cyprus, Czech Rep., Denmark, Estonia, France, Finland, Germany, Hungary, Ireland, Italy, Latvia, Lithuania, Luxembourg, Malta, Netherlands, Poland, Portugal, Romania, Slovakia, Slovenia, Spain, Sweden, UK		Greece	

Source Katz et al. (2011)

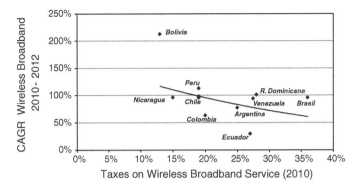

Fig. 4.15 Taxation versus adoption of data services. *Source* Katz et al. (2011)

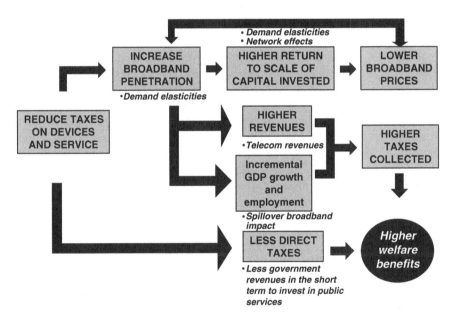

Fig. 4.16 Virtuous circle of tax reduction on broadband devices, equipment and services. *Source* Developed by the author

Pakistan—where the taxation approach runs counter to maximizing broadband adoption (Table 4.17).

In this context, taxation could have a detrimental effect on the public policy strategy aimed at deploying broadband. The impact of these different taxation approaches on total cost of ownership of broadband service varies widely. For example, in Mexico the impact of taxes on total cost of ownership of mobile broadband is 18.4 %, in South Africa it is 15.2 %, in Brazil it reaches 29.8 %, while in Bangladesh it is 54.8 %. On the other hand, in Malaysia, the effect of taxes on mobile broadband cost of ownership amounts to only 6.1 %.

Taxation appears to have an impact on the deployment of mobile broadband. For example, *ceteris paribus*, there may be some association between the very high level of taxes in Brazil and its very low penetration level of 3G handsets. On the other hand, Malaysia shows a low level of taxes and a high 3G penetration rate. Similarly, an inverse relationship appears to exist between tax burden and adoption of data services when measured by wireless data as percent of service revenues (see Fig. 4.15).

If taxes limit adoption of wireless broadband, it is relevant to ask what the ultimate impact of reduced penetration might have on economic growth. A reduction of taxation on broadband devices, equipment and services could have a significant economic benefit (see Fig. 4.16).

As Fig. 4.16 indicates a reduction of taxes on devices and service has a positive impact on broadband penetration as a result of the elasticities of demand (as discussed in Sect. 4.3). The increase in broadband penetration improves the number of households connected per households served (in fixed broadband) and the number of mobile broadband subscribers per infrastructure deployed. This increase in penetration enhances the return on the network capital invested. A higher return on capital allows the broadband service provider to lower prices, which in turn has a positive impact on penetration.

At the same time, an increase in broadband penetration has direct and indirect effects. On the direct side, it means an improvement in the revenues of broadband operators and ISPs. On the indirect side, it enhances the contribution of broadband to economic growth and employment. Both effects increase the taxable base, which in turn grows the collected taxes beyond the amount foregone by reducing taxes on broadband devices and services. This effect yields higher welfare benefits.

4.4.1 Programs Aimed at Reducing Taxation on Access Devices

Some countries have reached the conclusion that while foregoing tax collections in the short run, a tax reduction strategy can result in additional adoption of devices and broadband usage, and consequently enhanced economic benefits in the long run.

Case Study 4.23: Computer Purchase Program (Pakistan)

To promote economic growth and sustainability, Pakistan committed to a universal ICT and broadband service policy. The policy, launched in 2007 with public and private companies, stressed affordable voice and data services, increased broadband access, and telecenters development. Operator revenues, mobile access promotion charges, and spectrum auction proceeds funded the initiative.

Amongst other projects, the computer purchase program made home computers more affordable for students, government employees, and the military. For instance, all members of the Pakistani military qualify for a reduced-rate loan when purchasing a PC per a program established by Intel and the Ministry of Defense. Government employees and citizens requiring a computer for educational purposes qualify for a similar purchase program.

To foster a 1:1 e-learning environment, The Allama Iqbal Open University launched a computer purchase program offering all of its 700,000 students (including remote distance-learning students) below-market-rate loans for Intel-based laptop computers. The program, which was created by the Higher Education Commission and Intel, works with local banks to finance the loans. In 2011, Intel partnered with Meezan, an Islamic bank in Pakistan, to launch "Laptop Ease." In its first 4 months, the program provided 400 laptops to citizens with a 3–24 month repayment schedule. By 2012, the program aimed to increase this number to 250 laptops per month.

Sources

Pakistan Expands Broadband Connectivity and ICT Services to Bridge the Digital Divide. Rep. N.p.: Intel World Ahead, 2008. Print.

"Intel Offers Laptop Loans in Pakistan." Reuters, 19 Oct. 2011. Web. http://www.reuters.com/article/2011/10/19/intel-sharia-pakistan-idUSL3E 7LJ14920111019.

Case Study 4.24: Broadband Tax Relief (Malaysia)

In January 2010, the government of Malaysia began offering a 100 % capital expenditure tax allowance to operators investing in last-mile broadband equipment. This equipment and consumer access devices also qualified for import duty and sales tax exemptions. Further, the government offered an incentive to consumers, who received tax relief on the broadband subscription fee up to US$ 165. The country's Inland Revenue Board (IRB) announced in 2011 that these tax benefits extended to smartphone broadband use as well. While the actual devices did not qualify for the additional personal computer tax incentive, tablets such as the iPad did. The Malaysian Income Tax Act 1967 states that the cost of a personal computer—up to US$ 1,000—is deductible.

The broadband relief plan, which ran from 2010 to 2012, was an attempt by the Malaysian government to match the broadband use of neighboring countries in the region. At the time of its enactment, for instance, Singapore and Korea boasted respective penetration rates of 88 and 95 %, while Malaysia's hovered at 26 %.

The plan also made broadband access more affordable for 100,000 local university students, who qualified for a laptop with free broadband for US$ 16 per month for 2 years. This package was made possible by Telekom Malaysia.

Sources

Chung, Yee Sye. *Malaysia Broadband: A Leverage To National Growth.* Rep. SKMM, n.d. Web. http://www.myconvergence.com.my/main/images/ stories/SpecialEdition/pdf/MyConBumper_p13-17_MalaysianBB.pdf.

"Broadband Users Relieved over RM500 Tax Relief." *The Star Online.* N.p., 24 Oct. 2009. Web. http://thestar.com.my/news/story.asp?file=/2009/ 10/24/budget2010/4970057.

References

Atkinson, R.D., Correa, D.K., Hedlund, J.A.: Explaining International Broadband Leadership. http://archive.itif.org/index.php?id=142 (2008). Accessed 29 March 2014

Cadman, R., Dineen, C.: Price and income elasticity of demand for broadband subscriptions: a cross-sectional model of OECD countries. http://spcnetwork.eu/uploads/Broadband_ Elasticity_Paper_2008.pdf (2008). Accessed 22 March 2014

DANE.: Gran Encuesta Integrada de Hogares. http://www.dane.gov.co (2012). Accessed 2 April 2014

Dutz, M., Orszag, J., Willig, R.: The substantial consumer benefits of broadband connectivity for US households. Internet Innovation Alliance (2009)

Galperin, H.: Precios y calidad de la banda ancha en América Latina: Benchmarking y tendencias. Centro de Tecnología y Sociedad, Universidad de San Andrés. https://www.udesa. edu.ar/files/AdmTecySociedad/12_galperin.pdf (2012). Accessed 2 April 2014

Galperin, H., Ruzzier, C.A.: Price elasticity of demand for broadband: evidence from Latin America and the Caribbean. Telecom Policy **37**(6–7), 429–438 (2013). doi:10.1016/j.telpol. 2012.06.007

Goolsbee, A.: The value of broadband and the deadweight loss of taxing new technology. Contributions Econ. Anal. Policy **5**(1), 1–29 (2006)

Greenwood, J., Kopecky, K.: *Measuring the welfare gain from personal computers.* CRIW Workshop. NBER Summer Institute. retrieved: http://karenkopecky.net/ComputerSlides.pdf (2008)

GSR11.: Best Practice Guidelines on Regulatory Approaches to Advance the Deployment of Broadband, Encourage Innovation and Enable Digital Inclusion for All. Paper presented at 11th Global Symposium for Regulators, Armenia City, September 21–23, ITU (2011)

IBGE.: Pesquisa Nacional por Amostra de Domicílios. www.ibge.gov.br (2012). Accessed 2 April 2014

INDEC.: Encuesta Permanente de Hogares. www.indec.gov.ar (2012). Accessed 2 April 2014

INEC.: Encuesta Nacional de Empleo Desempleo y Subempleo ENEMDUR – Módulo "Tecnologías de la Información y Comunicaciones". ww.inec.gob.ec (2012). Accessed 2 April 2014

INEGI.: Encuesta Nacional de Ocupación y Empleo - Módulo sobre Disponibilidad y Uso de Tecnologías de la Información en los Hogares. www.inegi.org.mx (2012). Accessed 2 April 2014

InfoDev.: Mobile Usage at the Base of the Pyramid in South Africa. http://www.infodev.org/en/Publication.1193.html (2012). Accessed 2 April 2014

Katz, R.: El Presente y futuro de las telecomunicaciones en Costa Rica. Presentation to the IV Expo-Telecom, San Jose, Costa Rica. September 19. http://www.teleadvs.com/wp-content/uploads/Presentacion-Expo-Telecom-v.4.pdf (2012). Accessed 6 March 2014

Katz, R., Callorda, F.: Mobile Broadband at the bottom of the pyramid (report). http://gsma.com/newsroom/wp-content/uploads/2013/12/GSMA_LatAM_BOP_2013.pdf (2013). Accessed 2 April 2014

Katz, R., Flores-Roux, E., Mariscal, J.: The impact of taxation on the development of the mobile broadband sector (report). http://www.gsma.com/latinamerica/the-impact-of-taxation-on-the-development-of-the-mobile-broadband-sector (2011). Accessed 2 April 2014

Lee, S., Marcu, M., Lee, S.: An empirical analysis of fixed and mobile broadband diffusion. Inf. Econ. Policy **23**(3), 227–233 (2011)

Mitchell, J.: FCC Healthcare Connect Fund « Center for Telehealth and e-Health Law. http://ctel.org/expertise/fcc-healthcare-connect-fund/ (2013). Accessed 29 March 2014

NsrinJazani, A., Khatavakhotan, S.: A novel model for estimating bottom of the pyramid market size in IRAN based on inflation rate and income rate. IPEDR, vol. 13. IACSIT Press, Singapore (2011)

OECD.: Methodology for Constructing Wireless Broadband Price Baskets. OECD Digital Economy Papers, p. 205. OECD Publishing. http://dx.doi.org/10.1787/5k92wd5kw0nw-en (2012)

Prahalad, C.K.: The Fortune at the Bottom of the Pyramid. Wharton School Publishing, Philadelphia (2004)

Prahalad, C.K.: The fortune at the bottom of the pyramid: eradicating poverty through profits, p. 407. Wharton School, Upper Saddle River. ISBN 978-0-13-700927-5 (2010)

Rappoport, P., Kridel, D.J., Taylor, L.D., Alleman, J., Duffy-Deno, K.T.: Residential demand foraccess to the Internet. Int. Handb. Telecommun. Econ. **2**, 55–72 (2002)

Shah, Z.: The Income and Expenditure Pattern of Families at the Bottom of the Pyramid (2013)

Srinuan, P., Srinuan, C., Bohlin, E.: The Mobile Broadband and Fixed Broadband Battle in Swedish Market: Complementary or substitution? (2011)

Zager, M.: Broadband Transforms Rural Health Care. http://www.bbpmag.com/MuniPortal/EditorsChoice/0912editorschoice.php (2012)

Chapter 5
Developing Applications to Drive Broadband Demand

This chapter focuses on how to overcome one of the three dominant adoption obstacles to broadband diffusion among residential subscribers: lack of content relevance. The research reviewed in Chap. 2 indicated that, at higher penetration levels of broadband (beyond 20 %), price elasticity coefficients start to decline, signaling that affordability plays a smaller role in constraining diffusion. At that point, lack of relevant content remains the final obstacle to achieving mass adoption. While content relevance may also influence adoption at penetration rates below 20 %, price plays a stronger role in that situation.

Beyond pricing, demand stimulation also centers on enhancing broadband's value proposition. This is contingent upon offering applications and content that build service attractiveness. This chapter will review all potential initiatives to increase service attractiveness.

Section 5.1 of this chapter presents the multiple dimensions of content relevance, ranging from the linguistic to the cultural and applications dimensions. Once the context is established, the chapter will review options to tackle the content relevance obstacle (see Fig. 5.1).

Section 5.2 presents recommendations regarding the introduction of applications capable of building network effects, which could potentially stimulate broadband demand through viral diffusion processes. As expected, social networks, games, and other mobile-oriented applications have shown to be a powerful stimuli in promoting broadband adoption.

Section 5.3 reviews the launch of applications with high social and welfare impact, such as e-Government services, e-Health services, and financial services. Preliminary studies appear to indicate that, in addition to enhancing social inclusion, these applications provide additional stimuli to broadband adoption.

Section 5.4 deals with the introduction of applications and content generated within the local cultural context of the targeted population. Local customization can include linguistic characteristics as well as common cultural parameters.

R. L. Katz and T. A. Berry, *Driving Demand for Broadband Networks and Services*, Signals and Communication Technology, DOI: 10.1007/978-3-319-07197-8_5, © Springer International Publishing Switzerland 2014

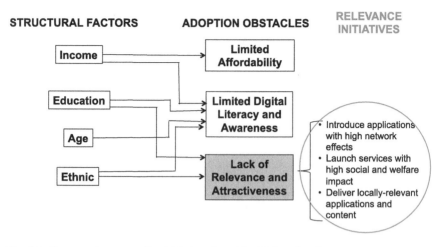

Fig. 5.1 Content relevance policy initiatives in residential broadband

5.1 The Multiple Dimensions of Content Relevance

Lack of interest or relevance appears consistently as one of the reasons cited by non-Internet users, regardless of their digital skills or income. This factor does not represent a barrier as such, and is linked to preferences and incentives that vary by community and from person to person. However, it represents a challenge for the development of public initiatives to address this aspect of the demand gap, since no single content solution can act as broadband demand stimuli across the board. Internet adoption studies reveal diverse mechanisms that come into play in the adoption decisions made by potential users, from which several possible public policy tools can be adopted.

Internet access, in itself, is of little value in the absence of so-called complementary goods that confer value to such access. A complementary good is one whose use is interrelated with the use of an associated or paired product (for example, DVD players and DVDs, computer hardware and software).[1] Examples of complementary dynamics in broadband adoption include applications and content that users value, and therefore should be attractive enough to encourage the purchase of the service. While early broadband adopters have the capability of rapidly identifying complementary products, late adopters (in other words, the target population of demand stimulation programs) require some help in exposing them to applications, services, and content whose complementary character adds value to the adoption of broadband. Three dimensions of complementary goods exist for broadband.

---

[1] While complementary goods are generally assessed in terms of its negative cross-elasticity (e.g. a reduction in the price of one leads to an increase in demand of the other), the concept is here used in terms of increased satisfaction derived from the consumption of the two products jointly (e.g. complements in consumption).

In the first place, the value of some applications increases as the number of users grows because of the use of the application to communicate or share information. This phenomenon is known as network effects or network externalities. At their most basic level, network effects exist in a service where its value to a user depends on the number of other users. Value is defined as the willingness to pay (or maximum amount that a consumer would pay) for acquiring the product. In general terms, willingness to pay in applications with network effects tends to increase as an "S" shaped function of number of users (see Fig. 5.2).

Research on broadband adoption indicates that network effects serve as a very powerful incentive to stimulate adoption. As mentioned above, applications designed for communication purposes between users have strong network effects. Four types of communications applications have these characteristics: point-to-point communications applications (e.g., email), social networking (e.g., Facebook, LinkedIn), content sharing platforms (e.g., Instagram), and matching networks (e.g., Monster.com). In point-to-point communications, access to a network facilitating communications is extremely valuable for users looking to communicate with friends, family, and members of a community they belong to. Social networking adds the value of self-representation to a community to the basic communications functionality. For example, Facebook allows users to communicate to network members via its Connect facility while presenting a personalized profile to the entire community. Content sharing platforms allow users to share either social (for example, pictures in Instagram) or business content (for example, documents in Dropbox). Finally, matching networks provide the capability to link participants with idiosyncratic needs and offers (for example, Monster.com links job seekers and companies offering employment). Strong single-sided or two-sided network effects power each of these communications applications. The net result is that the willingness to pay increases as a result of the application network effects and broadband prices decline. The combined effect becomes a stimulus for broadband adoption increases.

However, evidence also indicates that for certain groups of users, communication network effects may not offer enough incentive to purchase broadband services. In other words, for some consumers, the ability to communicate with friends, family, and community beyond voice telecommunications has little or no value. In this case, public initiatives should aim to provide high value-added applications, demonstrating tangible benefits to potential users in terms of saving time or money, or increasing welfare. Such is the case of the various e-government applications designed to help optimize the interaction of citizens with government, representing tangible benefits in terms of user access to different government services (health care, public administration, etc.).

In addition to the network effects and high social and welfare impact, a third driver of content relevance is embodied by language and cultural customization. This dimension is highly important in the case of users that belong to ethnic groups that communicate in languages not highly prevalent in the Internet or that do not find in the platform content more suited to their cultural background.

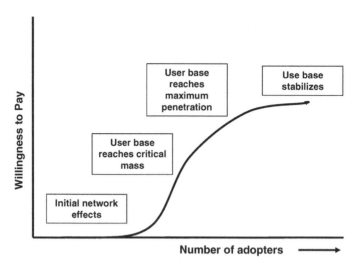

Fig. 5.2 Network effects growth. *Source* Developed by the authors

The following sections will provide examples of applications targeting each of these three relevance dimensions and review evidence of their impact in stimulating broadband demand.

5.2 Network Effects Applications Driving Broadband Demand

This section will review applications that stimulate broadband adoption by accelerating network effects. In each case, a description of the mechanisms linking network effect applications to broadband adoption is provided. To drive home the point, case studies of applications are introduced as examples.

5.2.1 Social Media Platforms

A social network serves as an Internet-based platform to articulate and make visible the users' current or past physical networks. In this sense, the platform allows users to construct a public profile within a bounded system, articulate a list of users with whom they share a connection, and view as well as navigate the user's list of connections and those made by others in the system.

The design of a social network follows a series of architectural choices. First, user profiles can be built around structured or flexible layouts and based on word descriptors or multimedia content. Second, the visibility of the profile can be open or user restricted. Third, the confirmation of network links (e.g., invitations) can be bidirectional or not. Fourth, the display of connections can be open or closed. Finally, the network can provide additional functionality, such as content sharing (e.g., pictures), built-in blogging, and mobile interface.

Social networks are the fastest growing Internet platform, with time spent having surpassed that of portals (see Fig. 5.3).

Social networks' initial value proposition was sharing user-generated content (web links, news stories, blog posts, notes, photos, videos, and events). However, very rapidly, they expanded into applications hosting platforms where developers would publish/upload applications, thereby enhancing allegiance to the site. For example, Facebook Platform comprises over 1 million developers who have uploaded over 350,000 applications. Games represent the largest category (over 13,000 applications being accessed by 418 million monthly active users) followed by the very distant lifestyle applications (4,800 applications), utilities (4,600 applications) and education (2,279 applications). Beyond user-generated content and applications hosting, social networks have also become a communications utility among platform users. The network becomes the vehicle by which point-to-point communications is conducted through services like Facebook Connect or LinkedIn Message. To these three dimensions (user-generated content, applications hosting and communications), social networks have in the past 2 years added the mobility dimension, which allows for "always-on" site connectivity.

Social networks are a worldwide phenomenon, in terms of their country of origin (see Table 5.1).

As Table 5.1 shows, beyond the platforms launched in the United States, China has become a leader in social network development.

Beyond the diversification of countries developing platforms, social networks initially developed within one particular country have grown into multinational platforms with a larger member base outside the country of origin. This is the case of Facebook, which has currently only 20 % of its members in North America, but a total of 26 % in Europe, 25 % in Asia, and 19 % in Latin America (see Table 5.2).

This deployment has been achieved through a sequence of internationalization and customization decisions:

- July 2004: Facebook launches in the United States
- January 2005: Facebook began to add international school networks
- December 2006: Facebook reaches nearly 2 million users in Canada and 1 million users in the United Kingdom
- January 2007: Facebook reaches 50 million active users worldwide
- January 2008: Facebook launches translated versions of the site in Spanish, French, German, and releases a translations application allowing users to translate the site into any language
- July 2008: Site reaches 90 million active users globally; users have begun to translate Facebook into dozens of languages

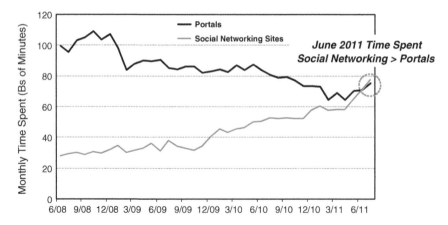

Fig. 5.3 USA: Monthly time spent, portals versus social networking sites, 6/08–7/11. *Source* comscore media metrix usa panel-only data

Table 5.1 Major Worldwide Social Networks (12/2012)

Network	Country HQ	Users (millions)
Facebook	US	937
Twitter	US	500
Qzone	China	480
Google+	US	400
Sina Welbo	China	300
Habbo	Finland	268
Renren	China	160
LinkedIn	US	160
Vkontakte	Russia	124
Bebo	US	117
Tagged	US	100

Source http://en.wikipedia.org/wiki/List_of_social_networking_websites

Social networks act as a powerful stimulus for broadband adoption. In general terms, countries appear to follow a sequential adoption process, whereby social network diffusion acts as an incentive for signing up on an Internet account through a ISP, lending itself over time to an increase in broadband adoption. The following three figures compile penetration time series for Facebook, Internet usage, fixed broadband and PCs for three emerging countries: Malaysia, Argentina, and South Africa (see Fig. 5.4).

In all three countries, the powerful network effects of social networks can be appreciated by witnessing the significant growth rate in Facebook penetration. Moreover, as mentioned above, the growth in Facebook penetration precedes by approximately 1 year the increase in fixed broadband adoption. This supports the notion that social networks act as a powerful incentive to adopt broadband. It

Table 5.2 Facebook: Breakdown of User Base by Country	Region	Number of monthly active users (million)	Total user base (%)
	Europe	243.2	26
	Asia	236.0	25
	North America	184.2	20
	South America	134.6	14
	Africa	48.3	5
	Central America /Mexico	47.0	5
	Middle East	22.8	2
	Oceania /Australia	14.6	2
	Caribbean	6.7	1
	Total	937.4	100

Source Internet World Stats 3Q2012 Report http://www.internetworldstats.com/facebook.htm

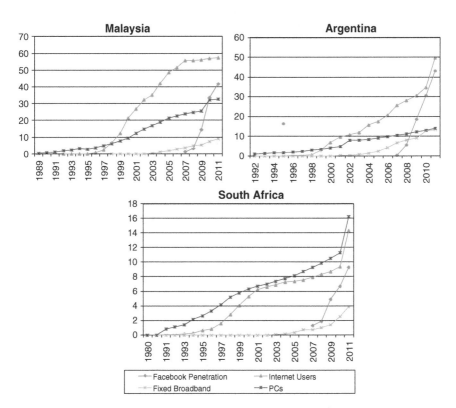

Fig. 5.4 Comparative penetration of social network, internet usage and fixed broadband. *Sources* Facebook statistics; World Bank; ITU; Telecom Advisory Services analysis

should be mentioned that the growth in fixed broadband penetration is mitigated by the fact that a portion of the Facebook "stimulating effect" is absorbed by mobile rather than fixed broadband.

In many emerging markets, carriers offer data plans, but with limitations. As a result, both Facebook and mobile carriers have begun experimenting with ways of offering free access to the social networking site.

In 2010, Facebook announced the launch of Facebook Zero, a service designed specifically for mobile phone use. Mainly targeting developing countries, the service promotes mobile Facebook access without data charges. When users go the Facebook Zero URL on their devices, they find a text-only version of the site, which carriers can offer free of charge. This "trimmed down" Facebook eliminates such data-intensive features as photos, but when users switch to the multimedia version of the site, they will then incur regular data use charges.

Table 5.3 lists the countries and carriers that partnered with Facebook for its initial Zero rollout.[2]

In other instances, carriers offer their own Facebook incentives to promote data use. As an example, Etisalat Egypt offers "Facebook+," granting unlimited Facebook access as part of its "Save More" mobile Internet plans.[3]

In Nigeria, Etisalat offered a Facebook promotion that allowed subscribers to download and use the mobile app for 90 days free of charge. Through the Etisalat Facebook Service, Facebook users could update statuses, post on walls, and upload photos by sending SMS messages to a specified number at a cost of N20 (US$ 0.13) per text (Table 5.4).

In addition to major private sector sites acting as broadband demand stimuli (Facebook, LinkedIn, Twitter, etc.), social networks can also be created ad hoc to fulfill specific purposes or solidify links across certain communities. For example, Ushahidi is a social platform developed in Haiti to facilitate networking among earthquake disaster victims. In other situations, existing platforms can be utilized to build links within a specific segment of a population. Such is the case in Russia, where the government utilizes Twitter to facilitate communication with its citizens. In these cases, ad hoc government sponsored networks can also play a role in stimulating broadband adoption.

---

[2] Wauters, Robin. "Facebook Launches Zero, A Text-Only Mobile Site For Carriers." *TechCrunch*. N.p., 16 Feb. 2010. Web. 11 Mar. 2013. http://techcrunch.com/2010/02/16/facebook-launches-zero-a-text-only-mobile-site-for-carriers/, and

Hopkins, Curt. "Facebook Zero Gives Free Access to Developing Countries - and Austria." *ReadWrite*. N.p., 18 May 2010. Web. 11 Mar. 2013. http://readwrite.com/2010/05/18/free_mobile_facebook_with_0facebookcom.

[3] "Mobile Internet." *Etisalat*. N.p., n.d. Web. 11 Mar. 2013. (http://etisalat.com.eg/etisalat/Etisalat_Portal_En/smartnet/smartmobile.htm).

"Etisalat Now Offers Free Access To Facebook For 90 Days." *TechLoy*. N.p., 5 Aug. 2011. Web. 11 Mar. 2013. (http://techloy.com/2011/08/05/etisalat-now-offers-free-access-to-facebook-for-90-days/).

Table 5.3 Facebook zero: countries and carrier rollout

Country or Territory	Operator	Country or Territory	Operator	Country or Territory	Operator
Anguilla	Digicel	El Salvador	Digicel	Nicaragua	Claro
Antigua & Barbuda	Digicel	Finland	Saunalahti	Pakistan	Mobilink
Aruba	Digicel	Greece	Vodafone	Palestine	Jawwal
Austria	tele.ring	Greece	WIND	Philippines	Sun Cellular
Barbados	Digicel	Guatemala	Claro	Romania	Vodafone
Belgium	BASE	Guinea Bissau	MTN	Rwanda	MTN
Benin	MTN	Honduras	Claro	Sri Lanka	Mobitel
Bermuda	Digicel	Hong Kong	3	St. Kitts	Digicel
Bolivia	ENTEL	Hungary	T-Mobile	Sudan	MTN
Bonaire	Digicel	India	Reliance	Swaziland	MTN
Brazil	TIM	India	Videocon	Tunisia	Orange
Bulgaria	M-Tel	Indonesia	3	Tunisia	Tunisiana
Cote d'Ivoire	MTN	Indonesia	AXIS	Turkey	Avea
Curacao	Digicel	Indonesia	Telkomsel	Turkey	Turkcell
Congo, Democratic Republic of the	Zain	Jamaica	Digicel	Turkey	Vodafone
Denmark	Telenor	Lithuania	Omnitel	UAE	du
Dominican Republic	Viva	Macedonia	ONE	Uganda	MTN
El Salvador	Claro	Madagascar	Orange		

Source Hopkins 2010

Table 5.4 Etisalat Egypt: Facebook+ offer for prepaid customers

Usage type	Plan name	Fees	MB included with the maximum speed	MB after reaching fair usage-throttled speed	Extra MB charges (L.E)	Speed reset fees	Facebook+
Light usage/irregular	Save more Weekly	7 LE/week	50 MB	Unlimited	Free	7 LE/50 MB	Free
Medium user	Save more Monthly 20	20 LE/month	110 MB			20 LE/ 110 MB	unlimited
High usage	Save more monthly 50	50 LE/month	500 MB			50 LE/ 500 MB	

Source Etisalat

Case Study 5.1: Ushahidi (Haiti)

Meaning "testimony" in Swahili, the nonprofit tech company Ushahidi develops free, open source software for information collection, visualization, and interactive mapping. Designed in 2008 for citizen journalists to map reports of violence and peace efforts via their computers or mobile devices following Kenya's election fallout, the website reached 45,000 users in Kenya alone. It subsequently grew into a global organization comprised of individuals ranging from human rights workers to software developers. The platform operates on the belief that the collection of crisis information from citizens offers a real-time and real-life glimpse, raising awareness and promoting efficient assistance and emergency deployment.

In 2010, a 7.0 magnitude earthquake in Haiti resulted in devastation and more than 230,000 deaths in some of the country's most populated areas. In response, the international community rallied to offer search and rescue missions and emergency assistance. While Haiti's system of information dissemination lacked the necessary sophistication, Ushahidi allowed victims and volunteers to coordinate through social media sites like blogs, Twitter, and Facebook as well as through mobile phone messages.

Reports submitted to Ushahidi included information about "about trapped persons, medical emergencies, and specific needs, such as food, water, and shelter" and were received and updated by volunteers in real time as well as by anyone who had an Internet connection regardless of their geographic location. Ushahidi deployed its platform within two hours of the earthquake, but quickly realized that it did not have the manpower to support the need for its services.

In response, Ushahidi's Director of Crises Mapping, Matrick Meier, reached out to volunteers at Tufts University to continue mapping the crisis live. The initial group of volunteers—all located in Meier's living room—covered social media to identify the information that needed to reach first responders on the ground. Utilizing Google Earth and OpenStreetMap, they tagged information based on GPS coordinates to upload it on the publicly accessible website Haiti.ushahidi.com. People from all over the world logged into the site to contribute what they learned via email or phone from relatives in Haiti.

Just days later, a volunteer team at Tulane collaborated with Frontlines MS, the United States State Department, and Digicel to create a system allowing Haitians to submit alerts via SMS text messages on their mobile phones. At the time, 85 % of Haitians had access to mobile phones and many of the towers were repaired immediately. The service was publicized over Twitter, radio announcements, and physical posters. The service received 1,000–2,000 SMS texts per day, but with a team of volunteers located in the U.S. and Canada, the system saw the messages conveyed to responders within 10 min. Volunteers translated more than 25,000 messages, emails, and social media communications, creating nearly 3,600 reports mapped on Ushahidi.

Sources

"About Us." *Ushahidi*. N.p., n.d. Web. (http://ushahidi.com/about-us).

Heinzelman, Jessica, and Carol Waters. *Crowdsourcing Crisis Information in Disaster- Affected Haiti*. Rep. United States Institute of Peace, Oct. 2010. Web. (http://www.usip.org/files/resources/SR252%20-%20Crowdsourcing %20Crisis%20Information%20in%20Disaster-Affected%20Haiti.pdf).

Case Study 5.2: Twitter (Russia)

In October 2010, the global media and ICT analytics firm ComScore released the results of the study it conducted earlier in the year, revealing that Russia ranks number one in terms of the world's heaviest social network usage. At this point in time, three-quarters of all Russians—or 34.5 million people—visited at least one social networking site in a one-month period. With the average user spending just shy of 10 h a month on such sites, Russians spend nearly twice as much time using social media than their global counterparts. Throughout the study, the top-ranked site, Vkontakte.ru, had 28 million visitors, though the number of Facebook users in the country grew 376 % in 2010.

To keep up with his constituents, then-president Dmitri Medvedev visited the San Francisco-based headquarters of Twitter, a global social networking and microblogging site. While there, he joined the site and sent out his first Tweet under the handle, KremlinRussia. From Russian, this message translated to "Hello everyone. I am now in Twitter, and this is my first message," but he also established an English-language version of KremlinRussia. This Tweet emphasized Medvedev's larger strategy of enhancing ICT innovation and investment back home and also demonstrated his push for a more transparent government. Within one year, Medvedev had created four verified accounts that attracted more than half a million followers.

Beyond showing his commitment to the aforementioned objectives, Medvedev and the Russian government used Twitter not only to share updates with the public, but also to "humanize" the president. The government's social media use now also includes other sites, such as i-Russia.ru, which the Presidential Commission for Modernization and Technological Development of Russia's Economy created for citizens to post comments and link government resources to their social network accounts.

Sources

Sniderman, Zachary. "How Governments Are Using Social Media for Better & for Worse." *Mashable*. N.p., 25 July 2011.

"Russia Has Most Engaged Social Networking Audience Worldwide—ComScore, Inc." *ComScore*. N.p., 20 Oct. 2010. Web. (http://www.com score.com/Insights/Press_Releases/2010/10/Russia_Has_Most_Engaged_Social_Networking_Audience_Worldwide).

"Dmitry Medvedev Visits Twitter HQ and Tweets." *The Telegraph*. N.p., 24 June 2010. Web. (http://www.telegraph.co.uk/technology/twitter/7852258/Dmitry-Medvedev-visits-Twitter-HQ-and-tweets.html).

Jagannathan, Malavika. "Russian Government Latest to Propose State-sponsored Social Network." *Herdict Blog*. Berkman Center for Internet and Society at Harvard University, 12 July 2012. Web. (http://blogs.law.harvard.edu/herdict/2012/07/12/russian-government-latest-to-propose-state-spon sored-social-network/).

5.2.2 Games and Gamification

Internet-based games, particularly the category known as social games, are applications strongly interlinked with social networking. As mentioned above, games represent the highest application category in Facebook Platform.

The importance of the games and social networking link is substantiated by the former's implicit value as a socialization tool. Research indicates that games allow users to overcome socialization obstacles by interacting with friends. Social games provide users with an opportunity to socialize when they lack content to post on their profiles. As such, they provide a structured way of interacting and keeping in touch with people. Additionally, the social game provides an excuse to re-engage with Facebook friends that are not that close to the user initiating contact.

For example, some of the more popular Internet-based games, such as Cityville or Farmville, present players with obstacles to build a business. Players can tackle obstacles by asking friends for help through a messaging system, or posting on their Facebook wall. Friends can reciprocate for help by sending free virtual gifts.

Along these lines, most social games tend to run inside social network platforms. This leads to a sequence of adoption processes leading to broadband: social game acts as an accelerator of social network usage, which, in turn, leads to broadband adoption.

An extension of game mechanics in nongame contexts is known as "gamification." Gamification is an approach utilized to enhance user engagement in Internet-based applications. Initial gamification techniques entailed providing reward points to users who share experiences on location-based platforms. For example, foursquare, a location-based platform rewards members with achievement badges to those users that become more active "checking in" with the platform. A more common "gamification" technique entails displaying in a platform a progress bar or other type of visual meter to indicate how close the user is to completing a task the platform operator is trying to encourage (e.g., a survey, a social network profile, a higher status in a loyalty program). Gamification techniques can be utilized in applications such as employee-training programs, primary education, wellness programs, and market research.

When it comes to utilizing gamification techniques in stimulating broadband demand, the best application could be through social media. Sequential causation would work as follows: gamification acts as an incentive of user engagement in social network platforms, which in turn, become a driver of broadband service acquisition.

Case Study 5.3: SuperBetter (United States)

At the TED Global 2012 conference, Jane McGonigal presented her findings on online gaming, a talk viewed online by nearly 1.5 million people globally. In the talk, armed with statistics, she insisted that online gaming "can be more effective than pharmaceuticals in treating clinical depression and that just 30 min a day is correlated with significant increases in happiness." She went on to assert that online games can help people healing from injuries and address such global problems as obesity and climate change.

In a similar talk from 2010—which reached an audience of more than 2.7 million viewers—she stated, "My goal for the next decade is to try to make it as easy to save the world in real life as it is to save the world in online games." Following a 2009 concussion that left her bedridden for months, McGonigal began viewing her pain as a game to cope, which led to her development of the online game SuperBetter. The game incorporates activities "designed to boost physical, mental, emotional, and social resilience," which ultimately helped to alleviate her depression and anxiety.

Working with doctors, psychologists, scientists, and medical researchers, she eventually made the game public, reaching users all over the world including cancer patients and sufferers of chronic pain. The game encourages positive emotions as well as a social connection, two contributing factors to physical and emotional health. McGonigal cites the "post-traumatic growth" theory, whereby positive changes in one's life can occur through the exposure to challenges—which can include illness or depression. Per its website, SuperBetter is not easy; it forces users to dedicate themselves to reaching various goals and milestones. It also offers a way to build physical, mental, emotional, and social strength and is customizable to each user. Users can access the game on their browsers and also through the Super-Better mobile application designed for iPhones and iPads.

Beyond the actual game, the SuperBetter website offers tips for users to keep a positive attitude and stick to their commitments. A user ultimately "wins" the game by achieving the "Epic Win," a real-life goal set by the user before beginning the game. Once achieved, the user can choose a new goal to begin a new game. In late 2011, Ohio State University Medical Research Center began conducting trials for the game to determine its effectiveness. The study will conclude in October 2014.

Sources

"Jane McGonigal: The Game That Can Give You 10 Extra Years of Life." *TED: Ideas worth Spreading*. N.p., July 2012. Web. (http://www.ted.com/talks/jane_mcgonigal_the_game_that_can_give_you_10_extra_years_of_life.html).

SuperBetter. Web. (https://www.superbetter.com/).

Kanani, Rahim. "Gaming for Social Change: An In-depth Interview with Jane McGonigal." *Forbes*. Forbes Magazine, 19 Sept. 2011. Web. (http://www.forbes.com/sites/rahimkanani/2011/09/19/gaming-for-social-change-an-in-depth-interview-with-jane-mcgonigal/).

"Clinical Trial of a Rehabilitation Game—SuperBetter." *ClinicalTrials.gov*. U.S. National Institutes of Health, 1 July 2013. Web. (http://clinicaltrials.gov/show/NCT01398566).

Case Study 5.4: EVOKE (Africa)

In 2010, the World Bank Institute launched its online game, EVOKE, to serve as a "ten-week crash course in changing the world." Designed for users around the globe, it particularly targeted the youth population in Africa. Ultimately, the game encourages users to tackle pressing social issues while providing them with the real-world skills necessary for social innovation. Anyone—regardless of age or location—can play EVOKE for free. Set in the year 2020, users follow "the efforts of a mysterious network of Africa's best problem-solvers." Players discover clues to solve the mystery each week while developing their own networks and work together to develop solutions to development challenges.

The game first began in March 2010 and ran until May 2010. This first round saw more than 4,000 players from over 120 countries. Players who successfully completed all ten challenges within the ten weeks were recognized as World Bank Institute Social Innovators—Class of 2010. The top players also qualified for mentorships with social innovators and business leaders as well as for scholarships to attend the EVOKE Summit, held in Washington, D.C. In that sense, winning the game translates into real-life rewards.

Per Robert Hawkins, a Senior Education Specialist at the World Bank Institute and the Executive Producer of the game, "*EVOKE helps players learn twenty first century skills to become the social innovators who shape the future." Explaining the game's name, Creative Director Jane McGonigal— who has made it her mission to create online games with social purpose—says, "When we evoke, we look for creative solutions and learn how to tackle the world's toughest problems with creativity, courage, resourcefulness and collaboration."*

The World Bank Institute—the World Bank division that focuses on learning—developed the game, which InfoDev and the Korean Trust Fund on ICT4D sponsored.

Equipe Evoke, the second round of EVOKE, launched in September 2012 in Portuguese and English and focused on Brazil. Users could play it on their computers or mobile phones.

Sources

"WBI Launches EVOKE: A Crash Course in Changing the World." *World Bank Institute (WBI)*. World Bank, 18 Feb. 2010. Web. (http://wbi.worldbank. org/wbi/news/2010/02/18/wbi-launches-evoke-crash-course-changing-world).

"EVOKE—the World Bank's Online Educational Game." *Education*. World Bank, n.d. Web. (http://web.worldbank.org/WBSITE/EXTERNAL/ TOPICS/EXTEDUCATION/0%2C%2CcontentMDK%3A22931151 ~ page PK%3A148956 ~ piPK%3A216618 ~ theSitePK%3A282386%2C00.html).

Case Study 5.5: Spent (United States)

Developed in early 2011 through a partnership with ad agency McKinney and Urban Ministries of Durham, the online game Spent shows users what it is like to live on only $1,000 a month in America. The game exposes the average player to real-life challenges, who must make the same decisions about money and resources that families across the country also face.

When a player first logs into the http://playspent.org website, he is greeted with a message: "Urban Ministries of Durham serves over 6,000 people every year. But you'd never need help, right?" Below this message is a link, where users can "Prove It" and accept the challenge. Once accepting the challenge, users are greeted by this question: "Over 14 million Americans are unemployed. Now imagine you're one of them. Your savings are gone. Can you survive? You've lost your house. You're a single parent. And you're down to your last $1,000. Can you make it through the month?" He can then choose "Find a Job" or "Exit." Once the game begins, players can choose from a selection of jobs, at which point the game breaks down the weekly take-home pay after deducting such expenses as taxes and the cost of supplies or uniforms. They then are faced with such decisions as whether they can afford health insurance or whether they should live in the city or the suburbs, taking into account the cost of housing and transportation. Other issues include the costs of childcare, the decision to join a union, and when to see a doctor. If a player hits a point where the only option to avoid homelessness is to seek help from a friend, he is directed to Facebook, where he can share the Spent link. At the end of the game, a player can also choose to donate to UMD.

By August 2011, users had played the game more than 1 million times in nearly 200 countries around the world. As a result of the game, users have gained a better understanding of poverty and homeless, which has led to an increase in the global dialog surrounding such issues. The creators of Spent have received feedback from people and organizations all over the world, each contributing a different anecdote to support the mission. As McKinney Chief Creative Officer Jonathan Cude states, the game shows and the feedback supports the idea that "poverty and homelessness can happen to anyone."

The Urban Ministries of Durham has used the game as an exercise to drive this point home to the community. McKinney and UMD also partnered to launch a petition to the U.S. Congress, asking "men and women of Congress to take 10 min from their debating to experience for themselves the challenges that more than 14 million Americans are facing" by playing the game.

Sources

Urban Ministries of Durham. *Spent, The Online Game About Surviving Poverty and Homelessness, Reaches Its Millionth Play and Invites Congress to Accept The Challenge.* N.p., 31 Aug. 2011. Web.

5.2.3 Mobile Broadband Applications

The accelerated diffusion of broadband access mobile devices has significantly increased the impact of mobile applications. In many cases, especially in emerging countries, mobile broadband may represent a substitute for fixed services in three types of situations: (1) the fixed service is not offered in the area where the user lives; (2) the quality of fixed services (for example, low speed) is less advantageous (or at least comparable to) than the available mobile service; or, (3) for economic or convenience reasons, the user opts to consolidate services and acquire only mobile broadband, which provides connectivity combined with mobility.

In the case of mobile applications, the adoption sequence is different from social media and games. In the prior two network effect groups, applications adoption precedes the purchase of a broadband subscription. Along those lines, the user becomes aware of the applications, begins using them, leading to an increase in the willingness to pay, which results in acquiring broadband services for home

Table 5.5 World: penetration of mobile broadband (2007–2013)[a]

	2007 (%)	2008 (%)	2009 (%)	2010 (%)	2011 (%)	2012 (%)	2013 (%)
Africa	0.11	0.59	0.98	1.56	2.9	5.4	9.47
North America	15.62	24.52	34.98	51.37	67.61	76.50	82.99
South America	0.58	1.25	3.01	6.82	14.41	24.57	41.16
Asia	1.18	1.78	3.04	5.77	10.53	16.51	24.60
Western Europe	2.12	5.21	11.79	20.70	35.15	52.88	70.92
Eastern Europe	0.42	1.30	4.77	8.41	13.99	22.34	36.56
Oceania	11.15	21.46	32.88	47.30	60.19	71.53	79.00
World	1.84	3.16	5.29	8.96	14.58	21.25	29.88
Developed	10.88	18.01	27.25	40.67	57.84	73.87	86.38
Developing	0.12	0.35	1.16	3.03	6.53	11.53	19.52

[a] Mobile broadband is defined as total subscribers of the following technologies: CDMA2000 1x EV-DO, CDMA2000 1x EV-DO rev. A, CDMA 2000 1x EV-DO rev. B, WCDMA HSPA, LTE, TD-LTE, AXGP, WiMAX, LTE advanced, TD-LTE advanced, WiMAX 2
Source GSMA intelligence

or individual access. In the case of mobile applications, the migration from an early generation (typically 2G feature phone access) to a 3G or 4G smartphone based device, which is an implicit adoption of broadband service, allows the user to gain access to mobile applications of which he/she was not previously aware. In sum, for social network and games, application use may precede fixed broadband adoption; in mobile broadband adoption, however, the purchase of service precedes application use.

In this context, mobile broadband users represent a "captive" market, ready to adopt mobile applications that enhance the value derived from wireless broadband. The installed base of wireless broadband users has been growing significantly, reaching an 86.38 % penetration in the developed world and 19.52 % in developing countries (see Table 5.5).

It should be considered, however, that, only a portion of this base corresponds to smartphones, devices whose screen format and keyboard are particularly suited to conduct Internet-based transactions. Nevertheless, the installed base of smartphones and data-only devices (USB modems, dongles) is also growing at a fairly rapid rate. In 2011, the most advanced regions in terms of shift to 3G and 3.5G smartphones are Western Europe (50 % of 3G and 3.5G base) and North America (49 % of base). In terms of the total installed base, North America has the highest smartphone share (40.3 %) followed by Western Europe (22.9 %). The rest of the world exhibits a smartphone share of 3G and 3.5G devices around 20 %. By 2015, smartphones in North America and Western Europe will reach 65 and 47 %, of mobile connections, followed by Eastern Europe (11.7 %), Latin America (13.1 %) and Asia Pacific (12.4 %).[4]

---

[4] Telecom Advisory Services. *Assessing the Worldwide Migration to Broadband Enabled Smartphones: a report to the ITU*, New York: November 23, 2011.

In the context of explosive adoption of broadband enabled devices, mobile applications serve to reinforce the awareness of broadband services. Governments can play a role in promoting the development of mobile broadband applications. Such is the experience of Canada, the United Kingdom and Singapore, reviewed below.

Case Study 5.6: Travel Smart (Canada)

In November 2012, Canada's Minister of State for Foreign Affairs announced the launch of its website, travel.gc.ca, which serves as a reference for international travel information and is fully accessible on mobile devices. In addition to the website, the federal government also released the free travel-related mobile app, Travel Smart, available for download on Blackberry, Android, and iPhone smartphones and tablets. The app addresses such issues as airline restrictions, security checkpoint wait times, passport and visa requirements, and letters of consent for traveling minors. All of this information is placed in one easily accessible portal.

Travel Smart connects with travelers through platforms like Twitter, Facebook, and SMS messages. It funnels information from 12 other federal agencies ranging from the Canada Revenue Agency to the Canadian Food Inspection Agency to the Public Health Agency of Canada.

The app is also designed for Canadians traveling outside the country who need to access safety information, learn about local laws and customs, and check on health conditions. They can register to the site before traveling so that the government will know where Canadian citizens are in the event of an earthquake or uprising abroad. Subscribers to the website's Twitter and Facebook feeds will receive travel updates and crisis information.

Beyond serving as a means for Canadians to access pertinent information more easily and quickly, such services allow the government to cut back on some of its more tangible expenditures. As the Department of Foreign Affairs cuts spending, it is closing international consulates and domestic trade offices. At the same time, the number of Canadians living abroad continues to increase year after year and citizens made 56 million trips out of the country in 2010 alone. An additional 2.8 million Canadians live abroad. As such, despite the need for cost cutting, there is also an increased need for available resources. In response, the government issued a statement saying that it must "find alternative and less costly ways to deliver routine services through the better use of technology."

Sources

MacKinnon, Leslie. "Government Touts New Mobile App for Smoother Foreign Travel." *CBC News*. N.p., 23 Nov. 2012. Web. (http://www.cbc.ca/news/politics/story/2012/11/23/pol-ablonczy-government-website-travel.html).

Case Study 5.7: Gov.sg Mobile App (Singapore)

Singapore's Ministry of Information, Communication and the Arts (MICA) released a mobile app in June 2010 to serve as a supplement to its website, Gov.sg. The "user-friendly" app allows smartphone users on the go to access government-related news stories and services such as event calendars, directories, and feedback forms. It serves as a seamless, one-stop platform for citizens to receive pertinent information from various agencies in a single integrated location. At the time, Singapore boasted some of the world's highest smartphone penetration rates in the world, in large part a result of its broadband and mobile infrastructure. With a 141 % mobile penetration rate, many citizens already had 3G and WiFi access, and this connectivity offered an "opportunity for the government to extend its range of services to better reach out to the people." Prior to the Gov.sg app, the government already utilized more than 300 mobile services.

The ministry initially launched the app for iPhones, but followed with an app designed for Android phones in November 2010 to meet the growing popularity of Google-based devices. With this second launch came an updated app design for the iPhone OS 4 that emphasized multi-tasking and built in a user-friendly interface.

Prior to the launch of the Android app, the Gov.sig iPhone app was downloaded 14,000 times in the first four months following its release. In its first week alone, the app ranked amongst the top three most popular App Store downloads in the "News/Free application" category. While the Singapore Government has already demonstrated its commitment to online government services, it hopes to encourage participation amongst the younger population through the development of mobile apps.

One year later, Infocomm Development Authority of Singapore partnered with NTUC's Employment and Employability Institute, SPRING Singapore and the Singapore Tourism Board to invest US$ 12 million to develop mobile applications for the service sectors. In response, the Ministry for Information, Communication, and the Arts stated that it "wants to harness opportunities presented by the mobile space and to develop mobile solutions for businesses." The development fund hopes to encourage growth in the mobile commerce and wireless sectors. To increase awareness and adoption of preexisting e-government services, the mobile website mGov@SG serves as a directory for these services and can be accessed by any phone with an Internet connection.

Sources

Singapore. Ministry of Information, Communications and the Arts. *Singapore Government Connects through New Mobile Application.* By Medha Lim. N.p., 29 Nov. 2010. Web. (http://app.mica.gov.sg/Default.aspx?tabid=36).

"Singapore Government Sets aside S$15 Million Fund for Mobile App Development." *Infocomm Investments*. N.p., 27 June 2011. Web. (http://www.infocomminvestments.com/2011/singapore-government-sets-aside-15-million-fund-for-mobile-app-development/).

"EGov2015 MGov@SG." *EGov Singapore*. Singapore Government, 28 Feb. 2013. Web. (http://www.egov.gov.sg/egov-programmes/programmes-by-citizens/mgov-sg).

Case Study 5.8: Government-developed Mobile Apps (United Kingdom)

By 2010, the British government had developed and released dozens of mobile applications serving a wide range of purposes, ranging from helping users quit smoking, to teaching them how to change a tire, to calculating mileage. One app, Jobcentre Plus, allowed users to search for open job listings across England, Scotland, and Wales. Following a 2010 Freedom of Information request, the government revealed that it had spent between US$ 15,000–$61,000 developing each app while spending US$ 143 million on website development and maintenance and US$ 52 million to support the necessary staff. The rollout of mobile apps could cut down on the website expenditures.

In 2012, multiple government agencies and service providers launched their own apps. The Socionical Crowd Sourcing app, for instance, was developed through a partnership between the London School of Economics and the UK police with funding from the EU. The app—available for download through the Apple App Store—allows users in London to download police-related information and offers catastrophe notifications. The police can monitor the movement of crowds through real time crowd sourcing technology.

Another government-launched mobile app now offers citizens a free online tax calculator to total the amount they owe and see how the government spends it. The app works on iPhones, iPads, and Android phones and is downloadable at the App Store and the HMRC website. The development of the app followed a report that revealed that more than half of taxpayers did not know the amount they paid in taxes. The government hoped that this new service would make the system more transparent.

By late-2012, Britain continued to push for the development of apps for government-related services. In November 2012, Prime Minister David Cameron announced that he was testing an app to aid in decision-making and day-to-day government affairs. The app has allowed Mr. Cameron to keep track of data as it pertains to areas such as jobs and housing in real time while monitoring political polling and social media. Following his approval, more government officials will have access to the app.

Sources

Schroeder, Stan. "UK Government Spends Thousands Developing IPhone Apps." *Mashable*. N.p., 6 July 2010. Web. (http://mashable.com/2010/07/06/uk-government-iphone-apps/).

Lee, Dave. "David Cameron Testing App to Aid Government Decisions." *BBC News*. N.p., 7 Nov. 2012. Web. (http://www.bbc.co.uk/news/technology-20240874?print=true).

Townsend, Christine. "City of London, UK Police Launches Smartphone App." *ConnectedCOPS*. N.p., 15 Aug. 2012. Web. (http://connectedcops.net/2012/08/15/city-london-uk-police-launch-smartphone-app/).

"Government Launches New Tax App." *Government Computing Network*. N.p., 29 May 2012. Web. (http://www.governmentcomputing.com/news/2012/may/29/tax-app-hm-treasury-hmrc-ipad-iphone-android).

5.3 Broadband-Enabled Services with High Social and Welfare Impact

Beyond applications with high network effects, some services with capability of having high social impact can act as an additional incentive to promote awareness of broadband's potential and further stimulate adoption. First, E-government services delivered via broadband can enhance citizen welfare by facilitating the conduct of transactions with public entities and providing access to government information. Secondly, broadband-enabled E-health services have the potential of enhancing the quality and reach of health care to citizens in isolated areas. Finally, financial services delivered via telecommunications, including broadband services, represent a powerful tool to promote financial inclusion of underprivileged segments of the population. Each category of services will be reviewed in turn, their potential to stimulate broadband adoption will be discussed, and examples of applications will be presented.

5.3.1 E-Government Services

Governments can enhance broadband demand by providing access to government services and broadband content online. A 2008 study by the Organization for Economic Co-operation and Development (OECD) notes the growing popularity of policy initiatives that foster more sophisticated online government services (OECD 2008). These initiatives include expanding secure government networks, putting administrative processes and documents online, supplying firms and citizens with more cost-effective ways to deal with the government (including once-only submission of data), and assigning firms and citizens a single number or identifier to conduct their relations with government. Government information available online in various countries includes legislation, regulations, litigation documents, reports, proposals, weather data, traffic reports, economic statistics, census reports, hearing schedules, applications for licenses and registrations, and even feeds from surveillance cameras. Some OECD countries even offer one-stop platforms for government procurement, bidding information, and so forth.

E-services that improve openness and access to democratic institutions are also becoming feasible as a result of increases in broadband transmission capacity. Examples include Internet broadcasts of parliamentary debates and agency meetings. Such applications allow citizens greater participation in the governance process. Applications for polling, voting, campaigning, and interacting with government officials can increase the demand for broadband services. In the United States, for example, two models of e-government citizen participation are emerging. One is a deliberative model where online dialog helps to inform policy making by encouraging citizens to scrutinize, discuss, and weigh competing values and policy options. The other is a consultative model that uses the speed and immediacy of broadband networks to enable citizens to share their opinions with the government in order to improve policy and administration. Actions to encourage citizen participation through e-government include the following:

- Connecting citizens to interactive government websites that encourage citizen feedback and participation in policy making, design, and innovation
- Encouraging citizens to participate in online dialog on topics such as health care and the economy
- Participating in government experiments with a variety of tools, including "wiki government," where citizens participate in peer review
- Educating citizens about their civic role and providing opportunities for them to interact with government agencies and officials using tools that fit individual or specific community needs
- Partnering with government officials and citizens to facilitate well-informed and productive discussions online
- Providing citizens the ability to create "my e-government" so they can personalize their interaction with government agencies and officials
- Creating "online town halls" to promote e-democracy for agenda setting and discussion of public issues as well as to promote accountability in the provision of public services.

A major goal of developing such services is to make government information more readily available as well as to increase transparency of government activities. The Netherlands is a leader in creating digital content and offering it via online government services (Atkinson, Correa, and Hedlund 2008, p. 39). In 2006, in an effort to support the development of broadband, the Dutch government decided to give all citizens a personalized web page—the personal Internet page—where they could access their government documents and social security information as well as apply for grants and licenses. In the United States, the U.S. government embraced e-government as an educational tool, particularly in providing online education programs for new immigrants seeking citizenship and for school support programs within the Department of Education. In Colombia, the 2010 Plan Vive Digital aspires to create a digital ecosystem by 2014 that would achieve several demand-related goals.

5.3.1.1 Types of e-Government Applications that Drive Demand

This section will examine the different types of broadband enabled e-government applications that have been launched in different countries. Government services and applications fall into the following broad categories: (a) making government information available, (b) conducting transactions with the government, and (c) participating in the political process.

In some cases, governments chose to implement discrete initiatives. Turkey's Land Registration and Cadaster Modernization Project serves as an example of the implementation of a discrete application to facilitate conducting transactions with the government. In other cases, governments opted to formulate holistic e-government policies and plans, such as the e-Government Act of the United States, the e-Governance Tools in the Netherlands, and the program Vive Digital in Colombia.

Case Study 5.9: Land Registration and Cadaster Modernization Project (Turkey)

An example of the Turkish government's expansion of e-government applications, the Cadaster Modernization Project was designed to make the land registry process more effective and efficient. The project took ownership of the renovation and updates of cadaster maps to create digital registry information, making this information available to the public, developing related policies, and improving services and resources.

While the land registration process will benefit from the initiative, it is the government's ultimate intent to improve its services while providing relevant information to the public and private sectors. In effect, it will raise awareness of the benefits of using online services for government-related matters, thus increasing the demand for broadband.

The project received funding approval in 2008 with an expected completion date of 2013. In total, it was expected to cost US$ 210 million, US$ 203 million to be loaned by the World Bank. Turkey's General Department of Cadaster and Registration oversaw its implementation.

Based on the World Bank's progress matrix, the project met many of its goals. The reduction in time for registrants to receive a response, for instance, decreased from 1 week to 2 h as a result of the online services. The number of agencies with online access to the services increased from 0 to 35 and the website processed approximately 6.5 million requests in 2012. Additional provisions were included to address such issues as technology training and policy drafting.

In 2013, the Bank approved a new procurement plan, which will have the project running through December 2014.

Sources

"Turkey Land Registration and Cadastre Modernization Project." *Projects & Operations.* The World Bank, n.d. Web. (http://www.worldbank.org/projects/P106284/turkey-land-registration-cadastre-modernization-project?lang=en).

"Procurement Plan Land Registry and Cadastre Modernization Project Revision 5.0." World Bank, 10 July 2013. Web. (http://www-wds.world bank.org/external/default/WDSContentServer/WDSP/IB/2013/07/26/00035 6161_20130726121249/Rendered/PDF/798020PROP0P100Box0379791B0 0PUBLIC0.pdf).

Case Study 5.10: The e-Government Act of 2002 (United States)

The United States' E-Government Act of 2002 worked with its Office of Management and Budget (OMB) to create an Office of Electronic Government and establish IT polices and guidelines to promote online government services. It designated a council to improve federal information resources. To ensure efficiency, the agency submitted an annual report to Congress conducting privacy assessments. Further, the Federal Information Security Management Act worked with OMB to "link information security with enterprise architecture." Amongst other responsibilities, it developed IT security standards, policies, and training for all agencies while reporting to various Congressional committees.

Ultimately, the e-Government Act aims to promote the effectiveness, efficiency, and quality of government service through online services while also allowing state and local governments to purchase updated technology. As one of the largest purchasers of technology, the government could benefit from reduced prices on IT hardware, software, and services. In 2007, the GSA signed a US$ 68 billion contract to provide 135 federal agencies in 191 countries with telecommunications services for 10 years at a price discounted up to 40 %.

Prior to the act, the federal government lagged behind the IT advances of the private sector. To achieve the same technological standards as the top performing private sector organizations, the Office of e-Government developed Internet-based government services, encouraging citizen and business interaction. These services aimed to increase participation while promoting government transparency and improving its efficiency.

The act budgeted US$ 345 million for 2003–2006 and stipulated that federal agencies report into the OMB regarding their progress in enhancing public participation and developing electronic services. That said, the agency felt that the time and resources required to develop the reports took away from the efforts spent toward the end goal.

In its 2012 Annual Report, The Government Accountability Office addressed concerns voiced by Congress, recommending a plan for the future and also highlighting its initiatives over the prior 10 years. Noteworthy accomplishments included increased availability of government information to citizens, a website for the public to comment on proposed regulations, and the creation of policies to protect citizens' personal information.

The report praised the integration of e-government into agency business and the resulting increased public participation, recognizing OMB's role in providing implementation leadership, raising awareness, and disseminating information. It did, however, concede that it needs to require more detailed reporting from the agencies in the future and should focus on reducing the burdens of this process.

Sources

Connecting America: The National Broadband Plan. Rep. Federal Communications Commission, n.d. Web. (http://download.broadband.gov/plan/national-broadband-plan.pdf).

"E-Government Act of 2002." U.S. Department of Education, 23 Aug. 2003. Web. (http://www2.ed.gov/policy/gen/leg/egov.html).

United States. Government Accountability Office. *Electronic Government Act: Agencies Have Implemented Most Provisions, but Key Areas of Attention Remain.* N.p., Sept. 2012. Web. (http://www.gao.gov/assets/650/648180.pdf).

Case Study 5.11: e-Governance Tools (Netherlands)

The Netherlands recognized early on the power of ICT tools to ease the administrative and bureaucratic burdens and improve the government services available to its citizens. The country placed high priority on citizen inclusion and participation, relying on extensive consultation when developing online programs and activities to ensure demand of such services.

In 2011, the government announced "i-NUP," an extension of the National Implementation Program (NUP), stating its plan to introduce further e-Governance measures through 2015. Prior to this point, many of the country's ministries had already gone online, publishing pertinent information on their websites, archiving digital documents, and allowing citizens to book appointments with elected officials. The i-NUP program ultimately aims to create "a more accessible government," providing basic e-Services for citizens and creating e-Business opportunities.

The NUP, which ran until year-end 2010, represented a partnership between municipal and provincial governments, water boards, and the central government. Together they established the basic requisites for e-Government, promoting e-Access, e-Authentications, and e-Information. As a result, citizens can now perform such tasks as applying for permits, registering for unemployment, accessing social security information, and working with various departments. By the end of the program, more than 8 million citizens could reach their municipal government.

Responsible for the development of e-Government policy, the Ministry of the Interior and Kingdom Relations coordinates initiatives with various levels of government and works closely with the Government ICT Unit (ICTU) and the Government Shared Services for ICT (Logius). The Ministry of Economic Affairs, Agriculture, and Innovation oversees e-Government initiatives as they pertain to businesses.

The project includes many portals, including the "Rijksoverheid.nl" citizens' portal. Accessible through a personal DigiD login, the portal offers every citizen access to e-governance services including personal data and tracking. The portal also offers relevant government information and news.

In 2012, the country ranked second in the United Nations' global E-Government For The People study. This ranking came from the secure nature of digital government communications as well as the country's broadband connections and the strength of the iNUP program.

Sources

"OECD E-Government Studies: Netherlands." *Public Sector Innovation and E-government*. Organization for Economic Co-operation and Development, 2007. Web. (http://www.oecd.org/gov/publicsectorinnovationande-govern ment/oecde-governmentstudiesnetherlands.htm).

"E-Government in the Netherlands." *EGovernment Fact Sheets*. European Commission, Nov. 2011. Web. (http://www.epractice.eu/files/eGovern mentTheNetherlands.pdf).

"The Netherlands Second in World Rankings of Online Government Services." *Government.nl*. Government of the Netherlands, 30 Mar. 2012. Web. (http://www.government.nl/news/2012/03/29/the-netherlands-second-in-world-rankings-of-online-government-services.html).

Case Study 5.12: Plan Vive Digital (Colombia)

The Colombian government's Vive Digital, which aims to connect all citizens to the Internet by 2019, incorporates a nationally recognized plan to develop a digital ecosystem through projects that focus on infrastructure, services, applications, and user experience. By making more services available online, the plan promotes efficiency and transparency at all government levels while curbing corruption and encouraging citizen participation. To generate demand, Vive Digital projects allow citizens to conduct transactions and access emergency services and value-added services online. It also incorporates such applications as an online Congress, tele-working, and government intranet while aiming for "zero paper" in the central administration. By bringing more everyday services online, the government hopes to make broadband more attractive.

Various ministries offer individualized services online. The Ministry of Finance, for instance, will offer mobile banking services while the Ministry of Education promotes ICT awareness amongst schools, teachers, and students. The Social Protection Ministry has taken ownership of e-health programs and rights verification services. Cultural initiatives are also supported through a National Digital Library and the Ministry of Defense has taken a role in ramping up cyber security and cyber defense mechanisms.

Between 2010 and 2014, Vive Digital has a budget of US$ 12.5 billion in ICT investment, with US$ 2.2 billion provided by the central government. The program is recognized as an integral component of the larger "Prosperity for All" plan.

As a result of Vive Digital, the United Nations ranked Colombia second in the ULatin America region in its 2012 global e-government survey. By 2014, the set deadline for the project, 100 % of all national entities and 50 % of all local entities will offer online services. The plan will continue to utilize modern ICT technology to promote the development of affordable, online services and foster an environment to make broadband relevant to everyday life. Ultimately, it aims to increase national productivity and improve the standard of living in the country.

Sources

Colombia. Ministry of Information and Communication Technologies. *Vive Digital*. N.p., 2011. Web. (http://vivedigital.gov.co/files/Vive_Digital_2011_Ingles.pdf).

"United Nations E-Government Development Database." United Nations Public Administration Programme, 2013. Web. (http://unpan3.un.org/egovkb/datacenter/CountryView.aspx?reg=AMERICAS%20~%20South%20America).

Table 5.6 Relationship between broadband and e-government services

Effect I: e-Government stimulates broadband adoption		Effect II: Broadband adoption stimulates e-Government services usage	
Dependent variable: broadband penetration 2008–2010		Dependent Variable: Use of e-Government	
Independent variables: e-Government use, employment rate		Independent variables: Broadband penetration, employment rate	
	Total		Total
e-Government use (% of Internet use)	0.555***	Broadband penetration	0.099 **
	(0.2990396)		(0.0561522)
Employment Rate	0.003 ***	Employment rate	−0.0003
	(0.0006617)		(0.0003668)
R^2 adjusted	0.2654	R^2 adjusted	0.0549
Number of observations	69	Number of observations	69

Note ***, **, and * indicate statistical significance at 5, 10 and 15 %
Source Katz and Callorda (2011)

5.3.1.2 Expected Impact of E-Government Services on Broadband Demand

This section will provide evidence of the impact e-government applications have on broadband adoption. As in the case of network effects applications, e-Government services have also been found to stimulate broadband adoption. In a study conducted for the Colombia government, Katz and Callorda (2011) researched the causality linking e-Government services introduced by the program Vive Digital and broadband penetration. The researchers found that both variables were linked through a bi-directional causality (in other words, e-Government stimulates broadband adoption, but broadband adoption also acts as a stimuli of usage of e-Government services (see Fig. 5.5).

According to Effect I, the deployment of e-Government services acts as a complementary service to broadband, enhancing the consumer surplus and, therefore, the willingness to adopt broadband. On the other hand, broadband adoption due to ancillary network effect applications (e.g., social networking) leads to the identification of e-Government services to be adopted. The authors built two econometric models to test both relationships (see Table 5.6).

According to the regression measuring Effect I, an increase of 1 percentage point of e-Government users yields an increase of 0.55 % points in broadband penetration. This finding would imply that e-Government services represent a strong incentive to purchase broadband subscription for home usage.

The results of Effect II also indicate a positive inverse relationship between both variables. An increase of broadband penetration of 1 percentage point increases the use of e-Government services in 0.10 percentage points. This means that, when controlling for employment rate, growth in broadband penetration reduces the cost to access e-Government from the home, yielding an increase in growth.

Fig. 5.5 Interrelationship between broadband and e-Government services

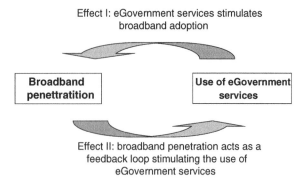

Effect I: eGovernment services stimulates broadband adoption

| Broadband penettratition | Use of eGovernment services |

Effect II: broadband penetration acts as a feedback loop stimulating the use of eGovernment services

The positive stimulation of government Internet applications on broadband applications was also found by Horrigan (2014) in a field research among low adopting households in the United States. According to this author, 50 % of surveyed families said that government agencies expect that they have home broadband services to access to e-Government applications.

5.3.2 e-Health Services

Broadband also contributes in an important way to the delivery of health care services. E-health involves a variety of services and tools provided by both the public and private sectors, including electronic health records and telemedicine. The communication technology is useful in tackling the economic challenges derived from delivering quality services, particularly in handling the relationship with patients, the provision of clinical care, and administration. In general terms, three priority areas can be identified where broadband can contribute to increase the efficiency of health care delivery. In the first place, the administration of health care information can be significantly improved on the basis of hospital information systems, digital archives of clinical histories, and treatments and digital libraries containing diagnostic images and records.

Secondly, patient-doctor/nurse communication can be conducted more efficiently by installing on-line registration, telemedicine (in areas such as tele-psychiatry, tele-cardiology, tele-radiology, and tele-surgery), remote monitoring, and the creation of community and social networks linking patients and health care professionals. The application of these technologies in the processes of both primary and specialized care has a positive impact not only on the quality of delivery but also on its flexibility and adaptation to patient needs. Hernandez, Leza and Ballot-Lena (2010) argue that citizens in rural areas, as well as those with limited mobility, may use e-health to access specialized care previously unavailable to them. For example, broadband capabilities are essential to medical evaluation and other medical applications that use imaging extensively. High-definition video consultations allow rural patients and immobile patients (for example, incarcerated individuals or nursing home residents) to be seen by specialists in a timely manner when urgent diagnosis is needed and the specialists are not able to travel to where the patients are located.

Thirdly, the communication among health care professionals through the deployment of social networks, distance learning programs, and video-conferencing leads to an improvement in skill level and the delivery of up-to-date information. In this way, health care professionals residing in remote areas can continue their training and updating by means of e-Learning platforms.

The contribution of these applications is far reaching. On a social dimension, research has identified an improvement in the quality of health care delivery, a decrease in time required for delivery of services, an increase in service quality, coupled with higher efficiency in the information exchange among professionals and the benefit of their continuing education. From an economic standpoint, the contribution of information and communication technologies to the health care sector results in better management of material and human resources, a reduction of care delivery and patient transportation costs, and the consequent reduction in information handling costs.

With the increased adoption of smartphones in low-income nations and the relative lack of wireline broadband penetration, mobile health (m-health) is establishing a new frontier in health care in those countries. Although basic voice and data connections are useful in improving health and medical care, broadband connectivity is necessary to realize the full potential of e-health and m-health services, particularly in rural communities.

On the other hand, broadband availability renders possible the development of new services as improvements in telemedicine and other e-health initiatives will rely on increasing bandwidth capacity, more storage and processing capabilities, and higher levels of security to protect patient information. As an example, Cape Verde has exploited growing broadband connectivity by connecting two of its hospitals to the Pediatric Hospital of Coimbra, Portugal (Favaro Melhem and Winter 2008). The telemedicine system supports remote consultations through video conferencing. One goal is to reduce the number of Cape Verdeans who must travel to Portugal for medical service. In addition to the Cape Verdean hospitals, two Angolan hospitals also are connected to the network and over 10,000 remote consultations have been carried out (CVTelecom 2010).

From an institutional standpoint, Horrigan (2014) concluded in a study cite above conducted among low broadband adopting families in the United States that 53 % of surveyed households said that health insurance companies expect them to have home Internet access.

5.3.2.1 Types of e-Health Applications that Drive Demand

This section will examine the different types of broadband enabled e-health applications that have been launched in different countries. There are three types of broadband-enabled e-Health applications: (1) transmission and remote access of

Table 5.7 World health organization mHealth initiatives

Category	Description
Communication between individuals and health services	• Health call centers/health care telephone help line • Emergency toll-free telephone services
Communication between health services and individuals	• Treatment compliance • Appointment reminders • Community mobilization • Awareness raising over health issues
Consultation between health care professionals	• Mobile telemedicine
Intersectoral communication in emergencies	• Emergencies
Health monitoring and surveillance	• Surveys via mobile phone • Surveillance • Patient monitoring
Access to information for health care professionals at point of care	• Information and decision support systems • Patient records

Source mHealth: New Horizons for Health through Mobile Technologies. Rep. World Health Organization, 2011. Web. 12 Mar. 2013. (http://www.who.int/goe/publications/goe_mhealth_web.pdf)

patient information, (2) patient-professional and inter-professional communications, and (3) mobile e-Health (adding the mobility functionality to the prior two applications categories). The three case studies reviewed below illustrate examples of all three categories. The Cape Verde experience illustrates categories 1 and 2, while the Indian and Ghana experience are examples of category 3.

A study published in 2011 by the World Health Organization (based on 2009 survey information) developed six categories of mHealth interactions. These initiatives included not only established mHealth programs, but also informal programs that used mobile technology for health-related communications and pilot programs undergoing evaluation (Table 5.7).

At the global level, the study found the most common initiatives to be:

• Health call centers/health- care telephone help lines (59 %)
• Emergency toll-free telephone services (55 %)
• Emergencies (54 %)
• Mobile telemedicine (49 %)

These four initiatives primarily rely on the voice feature of the phone. The occurrence of other functions of mobile health—such as raising awareness and downloading health information—will likely increase with increased access to mobile devices offering Internet connections and data storage capacities.

The adoption of specific mHealth initiatives also varied by country income (as measured by World Bank). The high-income countries reported a larger and more diverse range of initiatives, although all income groups reported that the health call center / help lines were the most frequently seen examples. High-income countries tended to utilize mobile devices to remind patients of appointments nearly twice as often as their low-income counterparts. The majority of these initiatives utilized voice, SMS, and Internet features of the phone. The service benefits not only individual patients and health care providers, but also the larger industry and economy. The United Kingdom and Northern Ireland, for instance, estimated that missed outpatient hospital appointments cost an annual US$ 1.2 billion.

In recent years, low and lower-middle income countries have looked to capitalize on the surge in mobile phone ownership to increase access to health care information. In doing so, they have addressed such issues as lack of quality health care professionals, high costs, and unreliable information. In the past, health call centers were limited to the high-income countries, where they have existed for decades. The present day ubiquity of mobile phones has closed this gap. In some instances, governments partner with local telecommunications operators to complement pre-existing health care services. In others, citizens pay a flat rate (e.g., $5 per month in Mexico) that allows for call center access and offers discounted physician services in instances that cannot be resolved over the phone.

Case Study 5.13: Integrated Telemedicine and e-Health Program (Cape Verde)

Developed in 2011 by the International Trust Fund (ITF) and the International Virtual e-Hospital (IVeH) Foundation, Cape Verde's Integrated Telemedicine and e-Health Program (ITEHP-CV) provides health-related online services. A nonprofit organization, IVeH introduces and implements telemedicine, telehealth, and virtual education within developing countries to create a sustainable telemedicine system.

With the country's Ministry of Health, the program establishes one main telemedicine center and nine local centers in pre-existing hospitals across the country. Beyond infrastructure and equipment installations, the initiative also trains medical professionals and establishes an electronic medical library. The project will "improve treatment outcomes, increase patient satisfaction, provide physicians' and other experts continuous education, reduce health care costs, and induce revolutionary transformation of the healthcare system in Cape Verde." Further, as citizens utilize these online services more frequently, they will see the value of high-speed Internet to their everyday lives

The program falls under the Development Cooperation Program, an agreement between the Cape Verde and Slovenia governments. Slovenia established the ITF to provide Bosnia and Herzegovina with mining-related funding, but now includes general global humanitarian projects. The Fund offered Cape Verde a US$ 29,000 research grant to identify telemedicine's potential as well as possible obstacles to the project's success.

Within months, two Cape Verde hospitals were linked to Portugal's Pediatric Hospital of Coimbra via a broadband connection. With video consultations, fewer citizens needed to travel outside of the country for medical services. By mid-2013, IVeH installed 10 telemedicine centers and trained more than half of all healthcare workers.

Sources

"Broadband Strategies Handbook." Ed. Tim Kelly and Carlo M. Rossotto. The World Bank, 2012. Web. (https://openknowledge.worldbank.org/handle/10986/6009).

"International Virtual E-Hospital Foundation Begins Cape Verde Telemedicine Project with ITF Enhancing Human Security." *International Virtual E-Hospital Foundation*. N.p., n.d. Web. (http://www.iveh.org/?id=82).

"Republic of Slovenia Supported Needs Assessment Mission in Cape Verde." ITF, 27 Sept. 2011. Web. (http://www.itf-fund.si/News/Republic_of_Slovenia_supported_needs_assessment_mission_in_Cape_Verde_369.aspx).

"Technical Installations of Telemedicine Completed on Nine Islands of Republic of Cape Verde." *International Virtual E-Hospital Foundation*. N.p., July 2013. Web. (http://iveh.org/?id=29&p=84).

Case Study 5.14: Remote Telemedicine Services (India)

In 2008, Ericsson partnered with Apollo Telemedicine Networking Foundation (ATNF) to establish an e-health program offering instant telemedicine services remotely over applications run on broadband networks. The service decreases the costs of healthcare while improving its quality and access, particularly for the rural population. Ericsson provides the necessary equipment while ATNF dispenses telemedicine advice instantly and remotely via customized applications. The HSPA network and mobile technology allows for increased access to quality healthcare and related services.

In India alone, more than one million citizens—particularly women and children—die as a result of limited healthcare. An additional 700 million rural citizens cannot access specialized services, as nearly all specialists reside in urban areas. ICT can extend the reach of quality healthcare beyond geographical boundaries without the same time and financial expenditures.

Prior, Ericsson received a license in 2007 to use 3G spectrum to demonstrate the impact it could have on e-governance, e-education, e-entertainment, and m-health through a pilot project known as "Gramjyoti." ATNF partnered with Ericsson to show the ways in which 3G can improve the transmission process of health information. Consultants at Apollo Group hospitals worked with patients in remote villages via text, audio, and video technology to "examine" patients with a web-based camera. Coupled with the patients' medical history and information, doctors then diagnoses 240 remote patients across 18 villages and 15 towns. Doctors and patients both reported high satisfaction.

Apollo hospitals conducted a similar pilot using EDGE technology. Utilizing the Hospital-on-Wheels (HoW) bus, volunteers took and developed X-rays, on board. Laptop software compressed the images and emailed them to doctors at the hospital. Doctors also conducted ultrasound examinations, which volunteers videotaped and then emailed to the hospital where internists could utilize the web cam to finish the exam.

These pilots have demonstrated the power of broadband to transform healthcare in India, but they have also shown the barriers to this progress, namely the lack of pre-existing infrastructure and the mindset of citizens in the targeted areas. Regardless, ATNF is India's largest telemedicine provider, and now has over 150 global centers.

Sources:

Bollineni, Raja. *Apollo Telemedicine Networking Foundation (ATNF)*. Rep. ACCESS Health International, 2011. Web. (http://healthmarketinnovations. org/sites/healthmarketinnovations.org/files/ACCESS_ISB_ATNF_Final_ Report.pdf).

Ericsson. Press Releases. *Ericsson and Apollo Hospitals to Bring Healthcare Access to Rural India*. N.p., 5 June 2008. Web. (http://www. ericsson.com/news/1225191).

"Apollo Telemedicine Networking Foundation (ATNF)." *The Center for Health Market Innovations*. N.p., 2014. Web. (http://healthmarket innovations.org/program/apollo-telemedicine-networking-foundation-atnf).

Case Study 5.15: Mobile Technology for Community Health (Ghana)

In 2009, Grameen Foundation launched its Mobile Technology for Community Health (MOTECH) initiative to reduce infant and maternal mortality rates. Healthcare providers can, for instance, collect and upload patient information to a centralized database. They can also use mobile technology to communicate with pregnant women, providing healthcare information and encouraging them to visit local medical centers for prenatal care while targeting women and newborns that could need their care through an automated patient-tracking service.

The Ghanaian government launched its Community-based Health Planning and Services (CHPS) Initiative in 2000, relocating urban clinic nurses to rural communities. While ambitious, the program ultimately did not take off because a) many of the women who needed healthcare could not access the village clinics, and b) nurses did not have the necessary knowledge and information to offer door-to-door services. Building on this initial program, MOTECH now delivers relevant information to the nurses via mobile phones and pregnant women can register for care by providing their phone numbers, location, and due dates. Once registered, they then receive SMS and/or voicemail updates relating to prenatal care, the location of clinics, and treatment recommendations. Some messages are personalized based on a woman's medical records and may, for instance, suggest certain vaccinations for her and her newborn child. In other cases, patients may send SMS questions to the nurses.

In collaboration with Columbia University's Mailman School of Public Health and the Ghana Health Service, MOTECH is made possible with a grant from the Bill & Melinda Gates Foundation. Johnson & Johnson, USAID, and the Norwegian government also provide project-specific funding. An extension of the project, the MOTECH Suite reflects a collaboration between international mHealth implementers, open-source solution providers, and funders to develop relevant software and establish best practices to guide future endeavors that will address health concerns using the MOTECH technology.

The program keeps costs low by running on mobile devices that citizens already own. The Gates Foundation covers the cost of messages. The country already boasts high mobile phone ownership and penetration, so device distribution and the associated costs associated are unnecessary. Small-scale field tests began in 2009 and the project officially launched in June 2010. In the pilots, the low literacy levels and variety of spoken languages posed the largest barrier. To overcome this challenge, MOTECH delivered either pre-recorded voicemails or standard SMS messages in the local language based on patient preference.

Sources

"Mobile Technology for Community Health (MOTECH)." *Grameen Foundation*. N.p., n.d. Web. (http://www.grameenfoundation.org/what-we-do/technology/mobile-health).

   "MoTeCH: A Mobile Approach to Maternal Health Care." *MobileActive*. N.p., 10 Feb. 2010. Web. (http://mobileactive.org/motech-new-approach-health-care).

5.3.3 Broadband-Enabled Financial Services

Online banking has evolved considerably, with the Internet becoming an integral part of the delivery of banking services around the world. It is generally recognized that e-banking services can provide speedier, faster, and more reliable services to customers and thus also improve relationships with them. Although many types of Internet connections have the capability to support online banking (for example, some m-banking transactions are conducted with narrowband short message service, SMS, messages), high-speed broadband is essential for effective e-banking.

Broadband enabled financial services have become prevalent in the industrialized world. A 2007 study, for example, found that in the United States, banking online was performed by 66 % of households with a home broadband connection compared to 39 % of households with a narrowband connection (DuBravac 2007, p. 9). This phenomenon is also prevalent in the developing world. Standard Charter Bank (India), for instance, now reports that an estimated 50 % of banking transactions in the country now take place online.

Moreover, delivering financial services to low-income users or populations of isolated areas through e-banking can also offer the potential to decrease operational costs dramatically, but more importantly, facilitate financial inclusion. Adoption of financial services in the developing world remains fairly limited. As of December 2010, over 2.7 billion people across the world did not have access to financial services. In Africa in particular, only 20 % of families have bank accounts,[5] leaving the vast majority of the continent "unbanked" with few prospects of ever creating viable savings. The vast majority of this demographic live in rural areas, beyond the growing urban sprawl of capitals and large towns leaving them with little exposure to the banking industry. Traditional banks have had very little interest in involving this lower income population in financial services. Beyond the usual fees required to open a bank account in most sub-Saharan countries, which most of the population cannot afford anyway, minimum deposit requirements can be as high as 50 % of per capita GDP.[6]

This problem is not limited to developing countries. For example, one in four U.S. households are defined as "unbanked" (not having a traditional checking account) or "underbanked" (forced to rely on other methods of money payment and management).[7] Additionally, FDIC studies in the United States show that amongst the 25 % of households that are underbanked or unbanked, the majority comes from a minority background. These households still need a method of paying bills and storing funds. In the past, when something needs to be paid for, cash was the first option. Traditionally, when cash couldn't be used money orders and money transfers were the solution for these individuals. However, today these methods are becoming more cumbersome and less generally accepted. Credit and debit cards are being used for more and more transactions, and online shopping is becoming the cheapest and most efficient way of procuring goods.

For these markets in particular, mobile money services have proved to be of particular importance. In countries such as Afghanistan, Bangladesh, Kenya, Indonesia, Pakistan, the Philippines, and South Africa, various forms of m-banking services are expanding the financial services frontier. These services allow users to make payments and remittances, access existing bank accounts, conduct financial

---

[5] Efam Dovi 2008 'Booting Domestic Savings in Africa', Vol.22, No.3.

[6] Ondiege.

[7] FDIC, "Survey of Unbanked and Underbanked Households", 2009.

Table 5.8 Categories of transactions in mobile payments

Category	Description	Comments
Pre-paid top-ups	Buying credits for pre-paid cards and phone service using mobile devices	Growth prospects high in developing countries
Money transfers	Peer-to-peer money transfers using proximity technology NFC (Near-Field Communications) or over the web	While accounting for only 21 % of mobile transactions, money transfers comprise 75 % of volume in the mobile payments market (due to higher dollar values vs. top-ups)
Merchandise purchase	Use of mobile phones as payment devices at the point-of-sale in retail locations utilizing NFC technology	Early days of adoption in developed countries (Japan and South Korea)
Ticketing	Mobile phones used as payment/proof-of-purchase at mass transit and public transport hubs (subways, taxis). Also includes sporting events, etc.	Market growing at nearly 100 % annually through 2014 and is expected to reach $3 billion in payments in 2014. Generally the lowest average transaction sizes versus other mobile payment categories
Bill pay	Mobile payments submitted via cell phone apps or via the web	One of the highest adoption rates in developed markets given consumer familiarity with web-based bill payment

Source Oreizy (2011)

transactions, engage in commerce, and transfer balances. A number of diverse financial transactions are broadly referred to as "mobile payments." It is important to understand that these various transaction types, however, possess very different characteristics and are often at significantly different levels of technological maturity and adoption. Below is a summary overview of the key categories in the mobile payments market (Table 5.8).

5.3.3.1 Types of Financial Services Applications that Drive Demand

This section will examine the different types of broadband enabled financial services applications that have been launched in different countries. There are three models for mobile payment that have been either discussed or implemented in various countries:

- Cellular provider serves as intermediary in transaction and stores the value as seen in the M-Pesa model (Kenya)
- Cell phones serve as a mobile wallet and allow the user to use existing accounts, or to consolidate charges made onto the cell phone bill or an internal credit card account (DoCoMo in Japan)
- Cell phones serve as mobile wallet but require a credit card / debit card to remit payments to (current US ISIS model)

Case Study 5.16: M-Pesa (Kenya)

In 2007, Vodafone's Kenyan subsidiary Safaricom launched the M-PESA system ("M" for mobile; Pesa, the Swahili word for money) for conducting electronic payments over mobile phones. Customers register at M-PESA retail stores, where they create electronic money accounts linked to their phone numbers and SIM cards. The stores convert customers' cash into an electronic value up to US$ 500, which can pay bills, purchase mobile airtime credit, and transfer funds. Safaricom compensates the retailers for each transaction. Registration and deposits are completely free. Customers must pay a US$ 0.40 flat fee for transfers and bill payments, US$ 0.33 for withdrawals, and US$ 0.01 for balance inquiries. Vodafone manages a server with all customer account information, but Safaricom manages these accounts. M-PESA accounts do not offer interest payments to customers, instead depositing this interest into Safaricom's not-for-profit trust fund.

In the 5 years prior to its launch, mobile penetration grew from 3 to 48 %, which largely contributed to M-PESA's success. M-PESA also succeeded due to the country's lack of financial services and the regulator's willingness to support Safaricom's business model.

Safaricom aimed to reach 1 million customers –17 % of its customer base at the time—within the first year. Prior to the nationwide launch, Safaricom conducted M-PESA pilots across the country, ensuring that 750 stores across all 69 districts of the country could handle the transactions. The emphasis on seamless customer service and brand recognition contributed to the system's popularity, which outpaced the initial 1 million-customer goal.

M-PESA demonstrated the power of mobile phone technology to bring financial services, to the unbanked, poor segment of the population. Many of these citizens already used mobile phones, allowing M-PESA to come to fruition without extensive infrastructure deployment or training costs. Further, with a usage-based revenue model, users, financial institutions, and mobile providers all experience benefits, eliminating the inherent discrimination as banks and providers targeted the more profitable customers. By connecting to this e-payment system, citizens can now save money in savings accounts, for instance, receive welfare disbursements, and send payments for services.

M-PESA has continued to grow rapidly. By 2012, the number of retailers outnumbered the country's bank branches. By 2013, more than 17 million Kenyans—or two-thirds of the population—had M-PESA accounts and the service handled 25 % of the country's GNP.

Sources

Mas, Ignacio, and Dan Radcliffe. "Mobile Payments Go Viral: M-PESA in Kenya." *Microfinance Gateway*. World Bank, Aug. 2010. Web. (http://www.microfinancegateway.org/p/site/m/template.rc/1.9.43376/).

O'Sullivan, Olivia. "The Invisible Bank: How Kenya Has Beaten the World in Mobile Money." *National Geographic Emerging Explorer*. National Geographic, 4 July 2012. Web. (http://newswatch.nationalgeographic.com/2012/07/04/the-invisible-bank-how-kenya-has-beaten-the-world-in-mobile-money/).

"Why Does Kenya Lead the World in Mobile Money?" *The Economist*. N.p., 27 May 2013. Web. (http://www.economist.com/blogs/economist-explains/2013/05/economist-explains-18).

Case Study 5.17: M-Paisa (Afghanistan)

Following the success of M-Pesa, in 2008, Telco Roshan launched its mobile money transfer program in Afghanistan, known as M-Paisa, in partnership with Vodafone Global Services. The two companies agreed to share profits, utilizing Vodafone's platform developed for M-Pesa. The word "paisa" means cash in Afghanistan's local languages, Dari and Pashto.

In 2010, the National Police in Afghanistan began receiving salaries through the M-Paisa system rather than in cash. Once the program was initiated, many of the policemen thought they had received a raise; in some instances the difference in pay equaled an increase of more than 33 % once higher-ranking officials could no longer take cuts for themselves from the cash payments. Beyond curbing corruption, the M-Paisa also discouraged policemen from defecting to the Taliban, who could previously offer higher payments for their services and loyalty. The funds are delivered to the policemen via an SMS and IVR system. The IVR system is particularly important given than 70 % of the country cannot read; these voicemails are available in Dari, Pashto, and English.

The Aga Khan Fund for Economic Development, Monaco Telecom International, and TeliaSonera back the provider. The Bill and Melinda Gates Foundation provided additional funding to M-Paisa through the US$ 12.5 mn grant it awarded to the GSMA trade association to provide mobile banking services to 20 million global citizens by 2012.

The service boosted Afghanistan's economy by curbing corruption and enabling small businesses and the Afghan people who previously had to deal with the risks associated with cash-only transactions. Beyond salary transfers, M-Paisa also enables airtime purchase, bill payment, and merchant services. At the time of the M-Paisa launch, less than 3 % of the population was banked and the financial sector was virtually nonexistent, in large part due to the prior 20 years of conflict. The high need for microfinance systems in the country drives the demand for such services and the Central Bank of Afghanistan has supported the development of M-Paisa.

By 2014, more than 1.2 million Afghans used M-Paisa.

Sources

Munford, Monty. "M-Paisa: Ending Afghan Corruption, One Text at A Time." *TechCrunch*. N.p., 17 Oct. 2010. Web. (http://techcrunch.com/2010/10/17/m-paisa-ending-afghan-corruption-one-text-at-a-time/).

M-Money Channel Distribution Case—Afghanistan. Rep. International Finance Corporation, 2010. Web.

Heinrich, Erik. "World How Afghanistan Is on the Leading Edge of a Tech Revolution Comments." *Time World*. N.p., 2 Mar. 2013. Web. (http://world.time.com/2013/03/02/how-afghanistan-is-on-the-leading-edge-of-a-tech-revolution/#ixzz2uN5XJe00).

Case Study 5.18: Splash Mobile Money (Sierra Leone)

In 2011, Splash Mobile Money Limited partnered with MoreMagic Solutions and the Guaranty Trust Bank to launch Sierra Leone's first mobile money service, Splash, to address the country's inadequate banking system. More than 150 certified agents throughout the country can register new users, converting their cash to "Splash Cash" credit on their phones. Users can then send or receive electronic funds, make purchases at participating retail vendors, or purchase prepaid airtime through a text message. After receiving Splash Cash, customers can go to any certified agent to convert the electronic money to cash.

At the time of Splash's launch, a mere 6 % of the Sierra Leone population had access to traditional banking services, making citizens more reliant on cash-only transactions. Splash customers can now use their mobile devices for financial transactions, eliminating high transaction costs, security, and limited transparency. Splash also eliminates rural citizens' burden of traveling long distances to a physical bank only to wait in line.

The company targets individuals and organizations alike. Sierra Leone's largest microfinance institution, for instance, sends loan disbursements and repayments through Splash. The country's largest satellite television provider uses Splash to accept payments from customers.

Splash received initial financing by way of equity funding from Manocap, the Soros Economic Development Fund, and through a grant from the Africa Enterprise Challenge Fund. Citizens' reluctance to trust electronic money initially posed a large challenge to the business model. Interestingly, the transparency of mobile-money transactions revealed the flaws and corruption in the traditional banking system, and this exposure resulted in resistance. Further, the low literacy rates posed a problem to the SMS platform.

Within the next 5 years, Splash hopes to expand beyond Sierra Leone and into at least five additional West African countries. It also plans to target the region's rural areas, thus bringing financial services to the unbanked population. As part of its growth plan, it aims to increase the number of agents as well as the opportunities for micro-franchising. By March 2013, mobile money transfer services in the country had a combined 1,000 agents, which has delayed their development and profitability.

Sources

Rudd, Melissa. "Making a Splash in Sierra Leone." African Business Review, 2 Nov. 2011. Web.

"From Harvard to Sierra Leone-CEO of Splash Mobile Money Finds His Path." *The AWP Network*. N.p., 28 Nov. 2012. Web. (http://awpnetwork. com/2012/11/28/from-harvard-to-sierra-leone-ceo-of-splash-mobile-money-finds-his-path/)

Sierra Leone–Mobile Money Transfer Market St. Rep. PHB Development, Mar. 2013. Web.(http://www.cordaid.org/media/publications/Sierra_Leone_Mobile_Money_Transfer_Market_Study_PHB_for_Cordaid_DEF. pdf).

Case Study 5.19: NTT DoCoMo DCMX Service (Japan)

In 2004, NTT DoCoMo and its eight regional subsidiaries launched the i-Mode Felica mobile wallet application for use on its i-mode 2G and 3G handsets. The devices could serve as train passes, debit cards, credit cards, and personal identification. Customers could purchase the handsets—which came equipped with Sony's Edy e-money system—at 9,000 shops throughout the country. 39 mobile providers offered m-wallet functions through the Felica service, available for download on the providers' websites.

In 2006, DoCoMo launched the "DCMX" consumer credit services through iD, its brand of mobile credit cards. Felica phone users could choose from two different plans based on their credit and purchasing behavior. The first plan, known as DCMX mini, offered a monthly credit line of US$ 100 and did not require a membership fee. Customers could use their phones for purchases without requiring separate identification. Payments were applied directly to the monthly phone bill. The other plan began at US$ 2000 and offered cash advances, requiring a four-digit password for purchases over US$ 100. This plan included a US$ 13 inactivity fee when the card wasn't used within 12 months and customers earned points redeemable for products and services offered by DoCoMo and its partners. Customers also received a Visa or MasterCard plastic credit card for use in stores that did not have readers.

By 2009, more than 30 million DoCoMo customers had purchased the Felica-enabled handsets, available for use on more than 420,000 readers through the country's retailers, restaurants, convenience stores, and taxis. Of these customers, 10 million regularly used the credit card service, reflecting the world's largest mobile payment market. In 2011, the company announced plans that the new models would support not only the FeliCa technology, but also the international standard N.F.C.

By 2013, Japan had six cashless payment systems, many of which were supported by mobile phone technology. Only 29 % of all mobile phone users, however, had registered for the e-money service.

Sources

NTT DoCoMo. *NTT DoCoMo to Launch Revolutionary Mobile Wallet Service Useable with New I-mode Smart-Card Handsets*. N.p., 16 June 2004. Web. (http://www.nttdocomo.co.jp/english/info/media_center/pr/2004/001182.html).

Kolesnikov-Jessop, Sonia. "Mobile Wallet Gaining Currency." *New York Times*. N.p., 6 Sept. 2011. Web. (http://www.nytimes.com/2011/09/06/technology/mobile-wallet-gaining-currency.html).

"NTT DoCoMo to Launch DCMX Mobile Credit Services." *Wireless Watch Japan*. N.p., 4 Apr. 2006. Web. (http://wirelesswatch.jp/2006/04/04/ntt-docomo-to-launch-dcmx-mobile-credit-services/).

White, Peter. "NTT DoCoMo Hits 10 m Mobile Wallet Accounts." *Rethink Wireless*. N.p., 25 Aug. 2009. Web. (http://www.rethink-wireless.com/print.asp?article_id=1826).

Fitzpatrick, Michael. "Death of Cash? Maybe, but Not Quite Yet in Japan." *CNN Money*. CNN, Fortune, 20 Feb. 2013. Web. (http://tech.fortune.cnn.com/2013/02/20/death-of-cash-maybe-not-so-fast/).

5.3.3.2 Expected Impact on Broadband Demand

There is some initial evidence that successful communications-enabled financial services act as a powerful incentive for communications adoption. The example of M-Pesa is quite illustrative. Research by Jack and Suri (2010) conducted among M-Pesa nonadopters indicated that the primary reason indicated for nonusage was lack of access (either lack of agents or cellular coverage) (21 %), while other factors (safety, ease, cost, confidentiality) scored significantly lower. This would indicate that, once the network is in place, adoption would greatly increase.

5.4 Content to Drive Broadband Demand

As discussed above, users purchase broadband services and devices in order to gain access to the complementary services and content. In fact, for the population at large, the network infrastructure is less important on a day-to-day basis than the availability of relevant and useful online services and applications that allow them to access, create, and share content.

This section will present different policies that could be formulated to stimulate the development of broadband-enabled applications and services in local languages, leveraging local content. Examples will be provided of the more successful applications, and details regarding ways of stimulating local development will be included.

5.4.1 Local Content Promotion Policies

In order to increase demand for broadband services, citizens must first view the service as relevant. Without resources, information, and applications designed with local communities in mind, the demand for such services will only come from the segment of the population for whom the Internet was first developed: native English speakers. Even with the rise in Internet users in countries where English is not the first language, comparatively fewer websites written in other languages and characters exist. Further, the defining characteristics of a culture—such as "geographic location, religion, ethnicity, and area of interest"—shape an individual's interest in available content.

The Internet has offered citizens the chance to create and distribute their own content more quickly and cost-efficiently than ever before. As discussed in prior sections, it has also allowed them to come together in such instances as crowdsourcing or mass broadcasting.

At the same time, however, lack of access further stratifies various segments of the population. Lack of access is only exacerbated by lack of local content, but in recent years, many developing countries have taken charge of the promotion and development of such resources. These efforts have commenced both to increase access and also to build a new industry of digital content.

In Kenya, for instance, the government allocated a budget of nearly US$ 4 million in an effort to increase locally relevant digital content and software. By working directly with developers, it hopes not only to increase demand for broadband services, but also to increase revenue within the industry. In the case of Egypt, governments have worked to digitize pre-existing cultural content, in turn encouraging Internet use while also allowing more people to benefit from its resources. In the Middle East, the UAE and Qatar have both developed industries centered on the distribution of digital content within the region.

In many instances, the governments turn to international corporations and organizations for both implementation and financial assistance in these endeavors. Much of this help allows these initiatives to benefit from pre-developed best practices while also establishing international support for the budding industries. Increasing content in areas such as education and technology training will also serve to strengthen the countries' economies.

In addition to direct grants for the production of local content, governments can support the development of local content and applications in other ways, such as the development of standardized keyboards, character sets, and character encoding. This type of indirect intervention would affect the content available by enabling users to create content in their own languages. Additionally, translation and standardization of operating systems into local languages can help to facilitate the development of local applications that are relevant and comprehensible to local users. Governments can also play an important role in developing local content and local applications by directly creating local content and local applications in the form of e-government applications, as described above.

Some forms of user-generated content, such as YouTube videos, face fewer barriers to expression, as the speaker is recorded directly in his or her own language. YouTube has launched a localization system, where YouTube is available in 31 local versions as well as a worldwide version. This helps to overcome some of the barriers to content reaching a possible community of interest, but not entirely, as content generated in languages other than those used in the 31 local versions or the worldwide version may encounter barriers to reaching an audience.

It is likely that greater amounts of local content will continue to become available in the near term. For example, a website called d1 g.com is a platform in Arabic for sharing videos, photos, and audio, a forum, and a question and answer facility. Launched in 2007, d1 g.com is one of the Arab world's fastest-growing social media and content-sharing websites, with more than 13 million users and 4.8 million unique monthly visitors. It has 15 million videos and streams an extensive amount of Arabic videos—600 terabytes of data per month. Notably, nearly 100 % of d1 g. com's content is user generated, with a small amount produced in-house. d1 g.com became the most popular Arab social media site (after Facebook and Twitter) when a user created the "Egyptstreet" diwan during the Egyptian revolution. During that time, unique visitors rose from 3 to 5 million per month, and visits per month grew from 6 to 13 million.

Case Study 5.20: Tandaa Local Digital Content (Kenya)

The Kenyan ICT Board's Tandaa "promotes the creation and distribution of locally relevant digital content." Tandaa defines digital content as "any content that can be consumed from an electronic device," such as a personal computer, digital television, or mobile phone, and is distributed through the Internet. Its website offers users the option to stay connected through

Facebook and Twitter and features intellectual property resources. In 2010, the ICT Board designated US$ 3.7 million for Tandaa's endeavors.

Sponsored activities range from digital training to research to content digitization. In 2010, for instance, the Company Registry of State Law saw all of its records digitized, with more than 20 million pages and records from 1936 to 2010 scanned and available online. The private company DPH won the tender offered by the ICT Board to manage the project, which involved 215 staff members working 24 h a day to scan 450,000 pages daily. Later in the year, the Board awarded its Digital Content and Software Application Grant to various firms in the government and private sectors, choosing grant winners based on their plans to introduce such initiatives as an HIV and AIDS in the workplace e-Learning Course, a teacher's portal, and a digital museum.

In 2011, Tandaa hosted four workshops to promote the second round of the Digital Content Grant. At each workshop, attendees learned about intellectual property rights and the application process and received training on proposal submission. The number of applications rose by 68 % from the year prior and nearly 50 % of applications came from outside of Nairobi, attributed to the workshops and to the redesigned online application process. Following the submission of proposals, the top applicants attended a business plan training session at Strathmore University.

In early 2011, the ministry created the Creative Content Task Force to promote Kenya's Visual Effects sector, which at the time was hindered by its fragmentation. Beyond connecting the various sub-sectors, the task force also worked with sector players to enhance the industry while increasing access to resources and trade opportunities.

In 2012, the IBM Service Corp Team developed an industry roadmap with representatives from UNCTAD, identifying the segments that the government should target based on their levels of innovation and economic activity. These sectors included new media, publishing, and visual arts industries. The roadmap demonstrated that lack of recognition, insufficient education and training, intellecutaul property rights, lack of ICT, and government policy issues hindered potential development. It then went on to establish ways to improve these areas.

Ultimately, per its mission statement, the roadmap identifies "how the creative industry can contribute 10 % of Kenya's GDP by 2017."

Sources

"Tandaa Kenya." N.p., 2012. Web. (https://sites.google.com/a/ict.go.ke/tandaa/home).

Case Study 5.21: Center for Documentation of Cultural and Natural Heritage (Egypt)

As the name implies, Egypt's Center for Documentation of Cultural and Natural Heritage (CULTNAT) oversees the documentation of Egypt's cultural and natural heritage and the dissemination and promotion of such resources. Topics range from archaeology, to national history, to music and folklore. Such projects have included an initiative to increase public awareness of cultural and natural heritage through the use of new media and training for professionals related to the conservation of cultural and natural heritage.

The Center emphasizes the importance of placing resources online using the latest digitization technologies, allowing the world to view and benefit from them regardless of their geographic location. To promote access to the Eternal Egypt project, for instance, CULTNAT partnered with IBM to develop an interactive, multimedia website where users can take guided tours of the country's pyramids in English, French, Arabic, Italian, or Spanish. The website utilizes "interactive technologies, high-resolution imagery, animations, virtual environments, remote cameras, three-dimensional models and more." CULTNAT also offers public exhibitions and projections that are available in print and CD formats, such as the atlas series that features maps, locations, and descriptions of the country's archaeological sites. Other initiatives include the world's first nine-screen interactive projective system, CULTURAMA.

The center is located in Cairo and falls under the direction of Egypt's Ministry of Communication and Information Technology. Much of the outside financial support comes from UNESCO, though project-by-project partnerships with the international private sector also offer a source of funding.

CULTNAT's work has not gone unnoticed by the international community. In 2003, the center received a Special Mention at the World Summit Award (WSA) for its work in the e-Culture category. In 2010, CULTANT won the WSA in the m-Tourism category as a result of its mobile-app, CULTMOB, which offers a tool for locating archaeological sites via 2.5G and 3G mobile devices. Its "Archeological Map of Egypt" program received the Stockholm Challenge Award in 2003/2004 and an award from the Arab Federation for Libraries and Information for the digitization of the country's National Historical Archives project. China's AVICOM International Committee for Audiovisual and New Image and Sound Technologies recognized CULTNAT's work in producing educational films.
Sources

"The Center for Documentation of Cultural and Natural Heritage (CULT-NAT)." *Bibliotheca Alexandrina*. N.p., n.d. Web. (http://www.bibalex.org/researchcenters/cultnat_en.aspx).

"WSA-mobile Outstanding Regional Achievement Awards 2010." *World Summit Award*. N.p., 2010. Web. (http://www.wsa-mobile.org/regional).

"CULTNAT." *The Center for Documentation of Cultural and Natural Heritage*. N.p., n.d. Web. (http://cultnat.org/).

Case Study 5.22: TwoFour54 (United Arab Emirates)

The Abu Dhabi Government published a long-term plan in 2007 that detailed its strategy for diversifying its economy away from its dependence on fossil fuels. Known as the "Abu Dhabi Economic Vision 2030," it emphasized, among other factors, the importance of a strong private sector, a knowledge-based economy, and a transparent regulatory environment while also detailing plans for the development of certain industrial sectors. The media and entertainment industry received special attention, largely due to the projection that its regional growth would exceed 19 % annually. With comparatively low broadband penetration rates that were expected to grow at 12 %, the government also saw this investment as a way to boost connectivity while taking advantage of the region's 3G mobile take-up and rise in popularity of social media. Further, the publishing industry in the MENA region – unlike in the rest of the world—continued to see positive growth in revenues and demand.

Named for the geographic coordinates of Abu Dhabi—24°North by 54°East—twofour54 promotes the development of Arabic-language media and entertainment. Launched in 2008, the initiative serves to place Abu Dhabi at the forefront of the media content industry, covering such segments as film, broadcast, music, digital media, gaming, and publishing. The project's campus includes three main segments: a training academy (Tadreeb), state of the art production facilities (Intaj), and support (Ibitkar).

Tadreeb, the training academy, offers 200 bilingual, international-standard courses through three of the project's partners, BBC, Thomson Reuters, and the Thomson Foundation. The courses reflect market demands and are offered either on campus or at client offices. Intaj offers such services as state-of-the-art HD production, post-production, media asset management and broadcast facilities to international production companies, producers, and broadcasters.

Ibitkar supports ventures for Arab entrepreneurs and businesses needing funding for the start-up phase, business development, or operational support. Ibitkar's creative lab offers grants to Arabs in need of seed funding or development assistance.

These three pillars are supported by tawasol, which looks after organizations and individuals looking to join the twofour54 community. Encouraging education, investment, collaboration, and partnership, the project provides Arabic companies and organizations with the tools, support, and infrastructure necessary to create high quality content. It also works to establish a seamless licensing process while working with intellectual property concerns.

Twofour54 boasts an impressive roster of media business partners. Many of these partners—like CNN, HarperCollins, and National Geographic—hail from all over the world and are leaders in the media, television, and publishing industries. Because the campus is a free trade zone, companies working with twofour54 enjoy the benefits of a tax-free environment with 100 % foreign ownership. The project has helped many of these organizations to expand their business to include the Middle East region, a virtue of both its geographic location and its understanding of the Arabic culture and market.

Sources

"Twofour54 Abu Dhabi." Abu Dhabi Media Zone Authority, n.d. Web. (http://twofour54.com/en).

5.4.2 Digital Content Promotion Policies

Case Study 5.23: Digital Content Incubation Center (Qatar)

Qatar's Supreme Council of Information and Communication Technology (ictQATAR) offers incubation services to new services, encouraging digital content innovation and the use of ICT. The Council ultimately envisions these companies contributing to the availability of Arabic digital content. Its Digital Content Cluster (DCC) offers resources such as innovative platforms, access to international partners, business development services, training, and office space. The entrepreneurs, start-ups, and small online businesses with which DCC works have access to conference facilities, IT services, government resources, coaching services, and accounting and legal services. They also receive support from multinational organizations like IBM, SAP, and Houghton Mifflin Harcourt.

This project is one of the Qatari government's many initiatives designed to support a knowledge-based economy with less dependence on hydrocarbon revenues. Investment in the ICT industry is expected to lead to the economic transformation that will drive sustainable development. Recent ICT projects have included the implementation of ultra-fast networks, international submarine cables, and satellite technology. More than 360 government e-services are now available online and the telecom sector now has a second mobile and fixed line operator. These innovations and the country's growing youth population and high purchasing power have led to an increase in demand for ICT services, and much of the growth in available services has come as a result of private sector investment.

To work with the incubation center, aspiring entrepreneurs must first pass through three phases—the eligibility phase, the admission phase, and the business phase—prior to admission. Once admitted, qualifying individuals will have access to the aforementioned resource for a specified tenure period and will eventually "graduate." Companies working with the center have addressed such topics as gaming, e-commerce, and women's issues.

Sources

"About the Digital Content Incubation Center." *QITCOM 2012*. N.p., 2011. Web.

Safla, Scheherazade. "A Call for Arabic Digital Content." *TFOURME*. N.p., 18 Nov. 2012. Web. (http://tfour.me/2012/11/a-call-for-arabic-digital-content/).

"Digital content" is defined as the myriad of websites, applications, and services available to broadband users. It can be based on text, audio, video, or a combination. Much of the content available on websites today can be divided into three broad categories: (a) user generated, (b) proprietary or commercial, and (c) open source.

User-generated content is produced within Internet-based platforms where users function both as consumers and as producers of content. Along these lines, consumers interact with one another instead of only dealing with site operators in a top-down fashion. User-generated content includes blogs, wikis, podcasts, Twitter updates, YouTube videos, and Flickr photos. They can be produced within social and professional networks, as well as reputational systems. For example, in the last three months of 2011, Facebook users collectively uploaded more than 250 million daily photos, equivalent to 7.5 billion photos per month or 3,000 photos per second. The social media site stores more than 100 petabytes, or 100 million gigabytes of photos and videos alone (SEC 2012). These forms of social media help to drive broadband demand by engaging users and ensuring the local and personal relevance of content. Due to the "bottom-up" nature of social media, policy makers can support the development of such content by taking a more hands-off approach in regulating it. They can also promote such services by becoming active users of such applications and services; more and more government agencies and even politicians are realizing the value of such tools in reaching out to citizens (see the example of Twitter usage in Russia).

As opposed to copyrighted materials, open-source content is available free-of-charge. In addition, the source code is also freely available to allow anyone wanting to incorporate the content or application into new forms of media, such as in mashups. Open-source content has led to the creation of property rights systems that encourage collaboration by publishing source code and allowing other users to extend those applications and develop them further, with the provision that the result should also be governed by the same open-source property rights.

Case Study 5.24: Digital Content Policy (Colombia)

In 2011, Colombia's ICT Ministry developed its Digital Content Policy to increase domestic sales from US$ 70 million to US$ 200 million by 2014. By encouraging the development of smart phone applications, video games, and digital animations, policy makers hoped that the country's digital industry would serve as the regional leader and attract higher levels of foreign investment.

To enhance the resources available to the country's professionals, entrepreneurs, and digital producers, MINTIC hosted Colombia 3.0, a non-profit initiative spanning 3 days. The conference invited industry leaders to network, participate in workshops, and exchange ideas. In 2012, Colombia 3.0 v.2 emphasized technological innovation and an interactive community, with Colombia 3.0 serving as a single point of convergence for the international digital industry. The conference encouraged outside investment while fostering an environment conducive to tech startups.

In 2011, Colombia 3.0 hosted 22 conferences, which saw nearly 3,000 and an additional 56,000 virtual attendees. In 2012, this number jumped to 40 conferences, with 4,000 attendees and 70,000 virtual attendees, which led to Colombia 3.0 in 2013 and 2014.

Sources

"Colombia 3.0, First National Summit of Entrepreneurs and Representatives of the Digital Content Industry in Colombia." Ministerio De Tecnologias De La Informacion y Las Comunicaciones, 30 Aug. 2011. Web. (http://www.mintic. gov.co/index.php/mn-english-news/368-20110907colombia30english).

"Colombia 3.0." N.p., n.d. Web. (http://www.col30.co/index.php? option=com_content).

Case Study 5.25: Two Trillion and Twin Star Program (Taiwan)

By 2001, Taiwan became one of the world's largest producers of semi-conductors and LCD technology. To encourage the growth of this sector, the Taiwanese government launched its Two Trillion and Twin Star program, investing in both the semiconductor and digital content industries. As a "rising star" industry, the government hoped that digital content would contribute more than US$ 30 billion annually. While "Two Trillion" part of referred to semiconductor industry growth, "Two Stars" promoted the digital content industry by encouraging Taiwanese companies to develop Chinese-language software and content.

This program spurred several digital content initiatives reflecting a partnership between the government and the private sector. The National Science Council sponsored 9 initiatives, including the five-year National Digital Archives Program (NDAP), which digitized collections from the country's museums, libraries, and universities, covering such themes as anthropology, painting, Chinese classics, and maps. The public could then access this content online from anywhere in the world.

The Two Trillion and Twin Star program also trained digital professionals through the Digital Content Promotion Office. The Creative Industries Promotion Office designated 13 areas of interest, including digital games and entertainment, focusing on their potential to increase employment in the country. Similarly, the Development Fund Investment Plan for Digital Content, Software, and Cultural Creative Industries financially supported digital content projects. The fund totaled US$ 3 billion, 40 % of which came from the government. The Executive Yuan also offered tax breaks and incentives.

Taiwan's electronic publication and digital archives output value grew 45.23 % from 2010 to 2011, reaching US$ 2.5 billion, and the digital content industry as a whole is expected to create new job opportunities, employing more than 70,000 citizens. The industry's output value reached US$ 18 billion in 2010, and grew to US$ 20 billion in 2011. This figure is expected to top US$ 26 billion in 2013, in large part due to Taiwan's presence in the global mobile application industry. Many of these mobile apps work with electronic learning, entertainment, and gaming content. In 2011, the Industrial Development Bureau established the App Incubation Center, working with domestic companies to encourage development. The government hopes that such an initiative will bring the country to the head of the Chinese-language app creation industry, producing 20,000 apps a year.

Sources

"Taiwan's Two Trillion, Twin Star (T3S) Plan." *GLOCOM Platform*. Japanese Institute of Global Communications, 6 Sept. 2002. Web. (http://www.glocom.org/tech_reviews/geti/20020906_geti_s22/index.html).

Liu, Yu-li, and Eunice H. Wang. ".tw Taiwan." *Digital Review of Asia Pacific 2007-2008*. Ed. Felix Librero and Patricia B. Arinto. New Delhi, India: SAGE Publications, 2008. 304-31. Print.

"Taiwan's Digital Content Industry to Grow Larger on Mobile Apps." *China Post*. N.p., 17 July 2012. Web. (http://www.chinapost.com.tw/business/asia-taiwan/2012/07/17/347878/Taiwans-digital.htm).

References

Atkinson, R.D., Correa, D.K., Hedlund, J.A.: Explaining International Broadband Leadership (2008). http://archive.itif.org/index.php?id=142 Accessed: 23 March 2014

CVTelecom: CVTelecom reinforces bet on Telemedicine (2010) http://www.telecom.pt/InternetResource/PTSite/UK/Canais/Media/DestaquesHP/Highlights_2010/Telemedicine_medigraf_CVtelecom_E.htm Accessed: 23 Mar 2014

Dubravac, S.G.: Broadband in America: Access, Use and Outlook. Consumer Electronics Association, Arlington (2007)

Efam, D.: Boosting domestic savings in Africa (2008). http://www.un.org/africarenewal/magazine/october-008/boosting-domestic-savings-africa Accessed: 3 Mar 014

Favaro, E., Melhem, S., Winter, B.: E-Government in cape verde. In: Favaro, E. (ed.) Small States, Smart Solutions Improving Connectivity and Increasing the Effectiveness of Public Services, pp. 155–191. The World Bank, Washington (2008)

FDIC: FDIC National Survey of Unbanked and Underbanked Households. http://www.fdic.gov/householdsurvey// Accessed: 23 Mar 2014

Hernandez, J., Leza, D., Ballot-Lena, K.: ICT Regulation in the Digital Economy (2010). http://www.itu.int/ITU-D/treg/Events/Seminars/GSR/GSR10/documents/GSR10-paper2.pdf Accessed: 23 Mar 2014

Hopkins, C.: Facebook Zero Gives Free Access to Developing Countries—and Austria (2010) http://readwrite.com/2010/05/18/free_mobile_facebook_with_0facebookcom Accessed: 11 Mar 2013

Horrigan J.: The essentials of connectivity. The Comcast Technology Research & Development Fund, Virginia (2014)

Jack, W., Suri, T.: The Economics of M-PESA: An Update (2010)

Katz, R.L., Callorda, F.: Medición de Impacto del Plan Vive Digital en Colombia y de la Masificación de Internet en la Estrategia de Gobierno en Línea. Bogotá (2011)

SEC.: Registration Statement on Form S-1. http://www.sec.gov/Archives/edgar/data/1326801/000119312512034517/d287954ds1.htm Accessed: 30 Mar 2014

Oreizy, P.: Runtime software adaptation: framework, approaches, and styles. ICSE Companion '08. pp. 899–910. ACM, New York (2011)

Chapter 6
Launching Services to Drive Broadband Demand

As discussed above in Sect 6.4, broadband in itself is of little value in the absence of so-called complementary goods that confer value to such access. Examples of complementary dynamics in broadband adoption include applications and content that users value, and therefore should be attractive enough to encourage the purchase of the service. However, beyond applications and content, complementarity also exists with the services that enable access to the Internet as well as value-added features that broadband operators include with the broadband subscription and that meet specific quality guidelines. The availability of such services is an important factor that influences and possibly *drives* demand. This level of demand, of course, will be affected, as discussed above, by the attractiveness and affordability of the service offerings.

6.1 Internet

A broadband subscription provides a high-speed connection to the Internet. The way the subscription is provided can affect attractiveness and will depend on the technology and regulatory or business considerations. This includes whether the broadband subscription can be purchased alone or whether it requires a subscription to an underlying transport technology. For example, in the case of a digital subscriber line (DSL) broadband connection, a telephone line is required. Subscribers have typically been obligated to pay a monthly rental for the telephone line in addition to the broadband subscription even if they do not use the telephone line for anything else but broadband. This adds to costs and may require an extra bill, discouraging users from taking up the service. Some operators include the telephone line with the broadband subscription, so there is no separate bill. In a few countries, the cost of the physical broadband connection is billed separately from Internet access. In other words, the user needs to pay one bill for a broadband connection and another bill for Internet access.

R. L. Katz and T. A. Berry, *Driving Demand for Broadband Networks and Services*,   257
Signals and Communication Technology, DOI: 10.1007/978-3-319-07197-8_6,
© Springer International Publishing Switzerland 2014

Table 6.1 Time required to download content at different service speeds

Content	56 kbps	256 kbps	2 Mbps	40 Mbps	100 Mbps
Google home page	23 s	5 s	0.64 s	0.03 s	0.01 s
5 Mb song	12 min	3 min	20 s	1 s	0.4 s
20 Mb video clip	48 min	10 min	1 min	4 s	1.6 s
CD	28 h	6 h	47 min	2 min	56 s
DVD	1 week	36 h	4.5 h	13 min	5 min

Source Broadband Commission Report, June 2011

Fig. 6.1 Type of user (segmented by download speed) in 2009 *Source* Adapted from the FCC Report "Broadband Performance: OBI Technical paper No. 4"

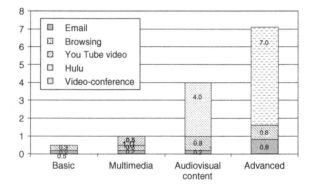

Several factors make a broadband subscription more or less attractive to potential users. One important factor is speed. Although some consider all "always-on" subscriptions of at least 256 Kbps to be broadband, in practice, speeds must be above a certain threshold to use desirable applications such as video viewing or gaming. It is important to establish that download speed has a direct impact on the user experience. Table 6.1 presents the times required to download different types of content.

However, it is important to consider not all broadband users require fast download speeds. While a minimum speed is certainly desirable for all users to have a good experience, beyond a certain point, desired download speeds depend on the content being downloaded. In study conducted by the United States Federal Communications Commission, broadband users were segmented into four categories according to the content being downloaded (see Fig. 6.1).

As shown in the figure, a user that relies on email and accesses the Internet via a search engine would rarely require more than 512 Kbps. However, once the user begins to download audiovisual content (such as You Tube video clips), it might require 1 Mbps. In case downloading includes TV shows, the download speed required to have an appropriate experience increases to 4 Mbps. Obviously, if the user needs reach the levels od advanced videoconferencing, download speeds will reach 7 Mbps.

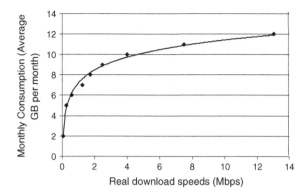

Fig. 6.2 United States: Download speeds and content volume in 2009 *Source* Adapted from the FCC report "Broadband Performance: OBI Technical paper No. 4"

In a typical developing country, one might observe that only 25 % of users could be characterized as "multimedia," thereby requiring only 1 Mbps. The problem is that the segmentation presented above is nothing but static. In fact, when assessing user behavior in terms of broadband consumption, one can observe that with more applications and content being available on the Internet, the higher the need for speed to confer an adequate experience. Furthermore, the faster speeds are offered, the more content is being downloaded. The same study conducted in the United States that segmented users by download speed observed the correlation between speed and content (see Fig. 6.2).

In summary, a variety of offers with different speeds provides more choice to the user. Other factors to consider are restrictions that the broadband providers may impose on capacity (for example, data or usage caps). Some operators distinguish between domestic and international use by having no cap or a higher cap for traffic to national sites and a low cap for access to sites hosted abroad. One issue with caps is that users often do not understand the relation between volume and their usage needs. Users can easily underestimate how much data they will use, particularly if they access a lot of video services or use peer-to-peer download services (some of which may run in the background). This makes it difficult for them to know which package to select when packages vary by data caps. Some operators cap usage through time rather than data volume (for example, monthly subscription of 20 h). These practices are more prevalent in mobile services, where "unlimited" offers (no download cap) are almost nonexistent.

Increasingly, governments are responding to data caps and "throttling" practices by requiring service providers to disclose their network management practices clearly, in order to protect consumers and improve the overall broadband. Regulators have also instituted other measures, such as monitoring quality of service and alerting users to sites where they can test their broadband connection for speed or throttling.

6.2 Voice

Voice telephony continues to be a popular service, although it represents a declining share of revenue for public telecommunication operators. A growing number of broadband operators offer voice over broadband (VoB) service, which is a managed service (unlike voice over Internet Protocol, VoIP, which is generally considered as an application running "over the top" of the public Internet and not directly managed by the network operator). VoB provides the same quality as a traditional fixed telephone and often provides other value-added features such as call waiting, voice mail, and speed dialing as well as the ability for users to monitor these features online via the provider's website. The price structure for VoB is often made attractive by including unlimited national calls for a flat rate or even including free national calls with the broadband service subscription. Since the service works through the broadband modem, users do not need to be connected to the Internet and do not even need a separate Internet subscription.

Several regulatory issues are related to VoB. The most basic is whether or not a country's laws and regulations allow it. Where VoB is legal, other regulatory considerations are often driven by the requirements placed on legacy wireline telephone networks. One is the requirement for users to be able to make emergency calls. Other regulatory requirements relating to consumers can include access for persons with disabilities and number portability. The latter can be influential in encouraging users to switch from traditional telephone services to VoB.

6.3 Video

IP networks allow video services to be provided over a variety of networks. This has allowed broadband operators to provide Internet Protocol television (IPTV) or video on demand (VoD) services. The ability to provide IPTV, VoD, or both can make operators' broadband services more attractive, especially when other features are included, such as access to special programming not available elsewhere. Television as a managed offering with a broadband subscription takes many forms. Some operators require IPTV to be bundled along with the broadband subscription, while others offer IPTV on a stand-alone basis.

Others have developed more extensive video service offerings, including BT in the United Kingdom, which offers its Vision service, which seamlessly integrates free-to-air digital television programs with a digital recorder and VoD feature. Some operators provide additional features such as radio programming and the ability to watch programming on computers, tablets, and mobile phones in addition to the traditional television set.

The ability to bundle television with broadband Internet service is often subject to technical and regulatory considerations. In the case of IPTV, users need to have a minimum bandwidth to use the service. Some countries require companies that provide television service to obtain permission or a specific type of license. Sometimes permission is required from local authorities. Conditions vary, but in general, television service is subject to a higher level of regulatory oversight than broadband service. Regulatory limitations have sometimes meant that operators can only provide delayed service rather than live programming, making their offer less attractive.

6.4 Bundling

IP-based technology and digitalization of media allow a single network to provide a variety of voice, data, and video services. The ability to offer multiple services has led operators to bundle services together. This often includes a price reduction in the total cost of the service (that is, the bundled prices is less than the cost of buying the same services individually) and the benefit of receiving just one bill. "Double play" refers to a combination of broadband Internet and some other service, "triple play" refers to the ability to provide three services, whereas "quadruple play" also includes mobile service.

Bundling offers can be attractive to consumers because of their lower costs and a single invoice. However, some consumers may only want one service from a provider and therefore need to have an "a la carte" option and not be obligated to purchase additional services. In any case, a service provider that is only allowed to provide Internet access is at a disadvantage versus converged operators since consumers are increasingly interested in receiving multiple types of communication services offered through bundles. Service bundles are currently offered as a promotion to limit customer disconnects through temporary discounts.

Chapter 7
Broadband, Firms, and Employment

In the last few years, spurred by the economic crisis, many governments around the world have implemented programs aimed at deploying broadband in order to stimulate employment growth (see Table 7.1).

How should one assess broadband effect on employment? Should it be considered an infrastructure development project necessary to build a platform to foster long-term economic growth? Or, alternatively, should it be conceived as a short-term job creation policy with only a speculative belief in its future employment multiplier effect? Obviously, either answer may be true. Nevertheless, it might be useful to understand how many jobs can be created by a broadband stimulus program, both in the short run (as a result of laying down fiber optics and erecting towers) and in the long run (as result of the potential innovations triggered by the deployment of a broadband highway that reaches all corners of the nation). In this context, it is important to differentiate between the two types of impact that broadband has on employment: jobs created to deploy the infrastructure, and network externalities that generate employment resulting from spillovers.

This chapter will review the evidence regarding the impact of broadband in terms of job creation. Differences will be made between the research focused on measuring the impact of broadband deployment programs (e.g., counter-cyclical impact of broadband network construction) and the spillover effect that broadband can have in terms of generating employment across the economy once it is being deployed.

7.1 Broadband and Job Creation

Broadband contribution to employment can be estimated both in terms of the direct impact resulting from network deployment (e.g., construction effect) and in terms of the indirect positive externalities derived from additional network coverage (e.g., network externalities). Each type of effect is comprised of three specific impacts (see Table 7.2).

R. L. Katz and T. A. Berry, *Driving Demand for Broadband Networks and Services*,
Signals and Communication Technology, DOI: 10.1007/978-3-319-07197-8_7,
© Springer International Publishing Switzerland 2014

Table 7.1 Counter-cyclical government programs

Country	Broadband focus
United States	Launched the $7.2 billion Broadband Stimulus program focused on providing and improving service to unserved and underserved areas
Australia	Government plans to spend A$ 11 billion of total A$ 43 billion required for construction of the National Broadband Network. Aims "to deliver broadband at speed of 100 Mbps to 90 % of Australian homes, schools and business through fiber-optic cables connected directly to buildings." The remaining 10 % "would get a wireless upgrade."
Germany	Government has announced a National Broadband Strategy with the objective to have nationwide capable broadband access (1 Mbps) no later than the end of 2010 and provide 75 % of German households access to a broadband connection of at least 50 Mbps by 2014 (estimated investment: Euros 36 billion)
Sweden	Broadband government promotion provides financial incentives to municipalities to fund 2/3 of total NGN investment (Euros 864 million)
Colombia	$290 million ($160 m for universal telephony, $50 m for ICT education, $30 m for Broadcasting, $30 m for computing education and $20 m for e-government)
Portugal	Government announced an 800-million-euro credit line for the rollout of NGAN. This is part of the first step in a 2.18-billion-euro plan to boost the country's economy. The investment should allow country to reach 50 % home broadband penetration by 2010. Aims to connect up to 1.5 MN homes and businesses to the new fiber networks and improve high-speed Internet, television, and voice services
Ireland	The government will invest 322 million in a National Broadband Scheme aimed at completing country coverage
Canada	Has relied on four programs to promote broadband development resulting in an overall investment of C$ 300 million. Focus on extension of broadband coverage to all underserved communities
Finland	Government funds one-third of the NGN project cost (S$ 130.73 m). The 7 years plan will "provide ultrafast speeds of at least one Mbps by 2010, with a ramp up to 100 Mbps by 2016. Includes households in rural areas"
New Zealand	Government funds S$ 458,12 m investment to boost fiber over the next 5 years
France	Estimated 10 years investment of US $13 billion. The plan will create 400 cyber bases in schools by 2012 and modernize schools that already have access. 4 million households will have ultra-broadband through FTTH access. By the end of 2010, affordable broadband Internet will be available throughout the country
Japan	Approximate US $370 million project. "Broadband infrastructure rollout plan for the rural areas, in order to address the digital divide, and to enable broadband access for use by cable TV, telecenters, disaster prevention programs, etc"
Singapore	The government announced a US $14.5 bn (6 % of GDP) stimulus plan in 2009, with US $650 MN dedicated to funding the Intelligent Nation Master plan, which includes NGN initiative. Aims (a) to connect homes and offices to the country's ultra high- speed and pervasive Next Generation National Broadband Network by 2013 and (b) for 60 % of homes and offices to have access to this all-fiber network in 2 years

(continued)

Table 7.1 (continued)

Country	Broadband focus
Republic of Korea	Won US $25 bn, US $940 MN from government. "High-speed Internet services to be upgraded to 1 Gbps by 2012; existing communications networks to be enhanced to Internet Protocol (IP)-based systems; subscriber capacity on 3G broadband services to be increased to 40 million." Expected to create 120,000 jobs
Spain	US $118 MN to be spent on infrastructure measures. Aims to expand broadband in rural and isolated areas by focusing on centers with dispersed populations and extending the reach of trunk fiber-optics networks

Source Compiled by the author, and Qiang, Christine Z. Broadband Infrastructure Investment in Stimulus Packages: Relevance for Developing Countries. Rep. N.p., 2010. Web. 12 Mar. 2013. http://siteresources. worldbank.org/EXTINFORMATIONANDCOMMUNICATIONANDTECHNOLOGIES/Resources/282822-1208273252769/Broadband_Investment_in_Stimulus_Packages.pdf

The following section will review the research conducted to date in assessing each type of effect. In addition, the section will present evidence of these specific effects in the context of developing nations.

7.1.1 Job Creation Through Network Construction

Broadband network construction affects employment in three ways. In the first place, network construction requires the creation of direct jobs (such as telecommunications technicians, construction workers, and manufacturers of the required telecommunications equipment) to build the wireline and wireless network facilities. In addition, the deployment of broadband networks has an impact on indirect employment triggered by upstream buying of inputs required for broadband network construction (which creates employment in, for example, metal and electrical equipment manufacturing sectors). Finally, the household spending based on the income generated from the direct and indirect jobs created by network deployment induces employment throughout the economy.

Input-output tables help calculating the direct, indirect, and induced effects of broadband network construction on employment. The interrelationship of these three effects can be measured through multipliers, which estimate how one unit change on the input side (e.g., investment in network deployment) affects total employment change throughout the economy.

According to input-output economics, multipliers are of two types. Type I multipliers measure the direct and indirect effects (direct plus indirect divided by the direct effect), while Type II multipliers measure Type I effects plus induced effects (direct plus indirect plus induced divided by the direct effect).

Six national studies have estimated the impact of network construction on job creation: Crandall et al. (2003), Katz et al. (2008, 2009, 2010), Atkinson et al. (2009), Liebenau et al. (2009). They all relied on input-output matrices and assumed a given amount of capital investment:

Table 7.2 Impact of broadband on employment

Macro effects	Specific impact	Description	Sectoral impact
Construction effects	Direct jobs	• Employment generated in the short term in the course of deployment of network facilities	• Telecommunications technicians • Construction workers • Civil and RF engineers
	Indirect jobs	• Employment generated in the short term in industries supplying inputs to network deployment sectors	• Metal products workers • Electrical equipment workers • Professional Services
	Induced jobs	• Employment created by household spending based on the income earned from the direct and indirect effects	• Consumer durables • Retail trade • Consumer services
Network externalities	Productivity	• Improvement of productivity as a result of the adoption of more efficient business processes enabled by broadband	• Marketing of excess inventories • Optimization of supply chains
	Innovation	• Acceleration of innovation resulting from the introduction of new broadband-enabled applications and services	• New applications and services (telemedicine, Internet search, e-commerce, online education, VOD and social networking) • New forms of commerce and financial intermediation
	Outsourcing	• Attract employment from other regions as a result of the ability to process information and provide services remotely	• Outsourcing of services • Virtual call centers • Core economic development clusters

- United States: US $63 billion needed to reach ubiquitous broadband service in this country (Crandall et al. 2003)
- Switzerland: CHF 13 billion to build a national multifiber network for Switzerland (Katz et al. (2008)
- United States: US $10 billion invested in broadband as a counter-cyclical stimulus (Atkinson et al. (2009)
- United States: US $6.4 billion invested in broadband as part of the Recovery Act voted by the US Congress in 2009 (Katz et al. 2009)
- United Kingdom: US $7.5 billion needed to complete broadband deployment in the United Kingdom Liebenau et al. (2009)
- Germany: 35.4 billion Euros to implement Germany's National Broadband Strategy as well as an Ultra-broadband Scenario targeted for 2020 (Katz et al. 2010)

Each study will be reviewed in turn

In 2003, Crandall et al. relied on the multiplier effects calculated by the Bureau of Economic Analysis, and concluded that US $63.6 billion of capital expenditures in broadband in the United States would trigger the creation of 61,000 jobs per annum. In addition, if investments were assigned to more advanced broadband platforms (such as VDSL, or FTTx), the cumulative effect of current and new generation of broadband would result in an increase of 140,000 new jobs per year. By adding both effects, the authors concluded that universal broadband adoption would trigger the creation of 1.2 million jobs, (546,000 in network deployment and 665,000 generated in upstream industries.)

In a similar vein, a study by Katz et al. (2008) estimated the impact of the deployment of a national FTTH network in Switzerland at a cost of CHF 13 billion. Using national input-output tables from Eurostat, the authors estimated that deployment of such a network could generate 114,000 jobs, 83,000 in direct and 31,000 in indirect employment. The study did not estimate induced employment.

Atkinson et al. (2009) also relied on input-output tables from the US Bureau of Economic Analysis to assess the employment impact of a $10 Billion investment in broadband infrastructure aimed at tackling the 2008 economic crisis. Their conclusion was that such an investment could create 64,000 direct jobs and 166,000 indirect and induced jobs.

In a replication of Atkinson et al. (2009), Katz et al. (2009) estimated the jobs that could be generated as a result of the grants to be disbursed by the broadband provisions of the American Recovery and Reinvestment Act, published in February 13, 2009. The study differentiated between jobs generated through capital spending in the form of grants allocated to unserved/underserved areas, and employment created as a result of network externalities caused by the deployment of such an infrastructure. Using input-output analysis to estimate the direct, indirect, and induced impact on jobs, the study found that approximately 127,000 jobs could be created over a 4 years period from network construction. According to the analysis, the investment of $6.390 billion[1] would generate 37,300 direct jobs over the course of the stimulus program (estimated to be 4 years). In addition, based on a Type I employment multiplier of 1.83, the bill could generate 31,000 indirect jobs. The split across sectors is presented in Table 7.3.

Once the Type I employment was calculated, the Type II effect was estimated. As mentioned above, the Type II effect refers to employment generated as a result of household spending derived from the Type I one. In this case, the Type II multiplier (direct + indirect + induced jobs/direct jobs) was 3.42. The combination of direct (37,300), indirect (31,000) and induced jobs (59,500) yielded a total employment impact of 127,800 jobs over a 4 years period. The average annual employment generation effect was estimated at 31,950 jobs per year.

---

[1] An estimate of funds dedicated primarily to broadband deployment, as opposed to ancillary activities such as broadband mapping.

Table 7.3 United States:
Type I employment effects of
broadband stimulus bill

Sectors		Employment
Direct employment	Electronic equipment	4,242
	Construction	26,218
	Communications	6,823
	Subtotal	37,283
Indirect employment	Distribution	9,167
	Other market/non-market services	8,841
	Transportation	1,536
	Electronic engineering	959
	Metal products	1,839
	Other	8,704
	Subtotal	31,046
Total type I employment		68,329
Type I multiplier (Direct + indirect)/direct		1.87

Source Adapted from Katz et al. (2009)

The investment multipliers for the United States broadband investment has been fairly consistent across research studies (see Table 7.4).

In other words, all three studies conducted to assess the employment contribution of a broadband construction program in the United States coincided in the magnitude of job creation. Differences only remained in the amount of capital to be invested.

In a similar approach to the US studies, Liebenau et al. (2009) calculated the labor impact of implementing the "Digital Britain Plan" in the United Kingdom. According to this study, the investment required to implement such a program would be US $7.5 billion, which could generate 211,000 jobs (comprising 76,500 direct and 134,500 indirect and induced).

Finally, Katz et al. (2010) estimated the employment impact of fulfilling the broadband deployment targets of Germany's National Broadband Strategy and an assumed Ultra-broadband Plan to be implemented by 2020. The estimated 35 billion Euros would generate 304,000 jobs over 5 years. The primary sector benefited in terms of job creation would be construction with 125,000, followed by telecommunications (28,400) and electronics equipment manufacturing (4,700). Total indirect jobs generated by sector interrelationships measured in the input/output matrix would be 71,000. The key sectors benefited from the indirect effects are distribution (10,700), other services (17,000), and metal products (3,200). Finally, household spending generated directly and indirectly, will result in 75,000 induced jobs. Based on these estimates, the Type I multiplier for employment would be 1.45 while the Type II would be 1.92.

Additionally, the implementation of the expected ultra-broadband evolution would generate 237,000 incremental jobs between 2015 and 2020. Similar to the breakdown reviewed above, this figure would comprise 123,000 in direct jobs, 55,000 indirect jobs and 59,000 in induced jobs.

Table 7.4 United States: comparative investment multipliers

Estimated investment (all $ numbers in millions)			Katz et al. (2009) $6,390	Atkinson et al. (2009) $10,000	Crandall et al. (2003) $63,600
Network construction	Employment	Direct effect	Jobs in equipment mfr, construction and telecoms	37,300	63,660	546,000
		Indirect effect	Jobs in other industries caused by direct spending	31,000	165,815	665,000
		Induced effects	Jobs in all industries	59,500		
		Total jobs	Type I and II	127,800	229,475	1,211,000
	Multipliers	Type I multiplier	(Direct + indirect) /direct	1.83		
		Type II multiplier	(Direct + indirect + induced) /direct	3.42	3.60	2.17

Source Adapted from Katz et al. (2009)

Type of impact	2014 national broadband strategy	2020 ultra-broadband	Total
Direct effect	158,000	123,000	281,000
Indirect effect	71,000	55,000	126,000
Induced effect	75,000	59,000	134,000
Total	304,000	237,000	541,000
Type I multiplier	1.45	1.45	
Type II multiplier	1.92	1.93	

Table 7.5 Germany: Total Employment impact of Broadband Network Construction

Source katz et al. (2010)

As in the other studies, the German case concluded that the labor-intensive nature of broadband deployment determines that the construction jobs to be created are significant and, despite the high-technology nature of the ultimate product, broadband is to be seen as economically meaningful as conventional infrastructure investment such as roads and bridges (Table 7.5).

To summarize, all studies that have relied on input-output analysis have calculated multipliers, which measure the total employment change throughout the economy resulting from the deployment of a broadband network. Beyond network construction (direct employment), broadband construction has two additional employment effects. Network deployment will result in indirect job creation (incremental employment generated by businesses selling to the sectors that are directly involved in network construction) and induced job creation (additional employment induced by household spending of the income earned from the direct and indirect effects) (see Table 7.6).

Cognizant that multipliers from one geographic region cannot be applied to another, it is useful to observe the summary results for the multipliers of the four input-output studies.

As mentioned above, according to the sector interrelationships depicted above, a European economy appears to have lower indirect effects than the US. Furthermore, the decomposition also indicates that a relatively important job creation effect occurs as a result of household spending based on the income earned from the direct and indirect effects. The induced effects are very large.

While input-output tables are reliable tools for predicting investment impact, two words of caution need to be given. First, input-output tables are static models reflecting the interrelationship between economic sectors at a certain point in time. Since those interactions may change, the matrices may lead us to overestimate or underestimate the impact of network construction. For example, if the electronic equipment industry is outsourcing jobs overseas at a fast pace, the employment impact of broadband deployment will diminish over time and part of the counter-cyclical investment will "leak" overseas. Second, it is critical to break down employment effects at the three levels estimated by the input-output table in order to gauge the true direct impact of broadband deployment. Having said that, all these effects have been codified and therefore, with the caveat of the static nature of input-output tables, we believe that the results are quite reliable (Table 7.7).

Table 7.6 Broadband construction impact on job creation

Country	Study	Objective	Results
United States	Crandall et al. (2003)	Estimate the employment impact of broadband deployment aimed at increasing household adoption from 60 to 95 %, requiring an investment of US $63.6 billion	• Creation of 140,000 jobs per year over 10 years • Total jobs: 1.2 million (including 546,000 for construction and 665,000 indirect)
	Atkinson et al. (2009)	Estimate the impact of a US $10 billion investment in broadband deployment	• Total jobs: 180,000 (including 64,000 direct and 116,000 indirect and induced
	Katz et al. (2009)	Estimate the impact of US $6.4 billion invested in broadband as part of the Recovery Act voted by the US Congress in 2009	• Total jobs: 127,000 jobs (comprising 37,300 direct and 31,000 indirect jobs)
Switzerland	Katz et al. (2008)	Estimate the impact of deploying a national broadband network requiring an investment of CHF 13 billion	• Total jobs: 114,000 over 4 years (including 83,000 direct and 31,000 indirect)
United Kingdom		Estimate the impact of investing US $7.5 billion to achieve the target of the *"Digital Britain"* Plan	Liebenau et al. (2009) • Total jobs: 211,000 (including 76,500 direct and 134,500 indirect and induced)
Germany	Katz et al. (2010)	Estimate the impact of investing 35.4 billion Euros to implement Germany's National Broadband Strategy as well as an Ultra-broadband Scenario targeted for 2020	• Total jobs: 541,000 (comprising 281,000 direct, 126,000 indirect, and 134,000 induced jobs

Source compiled by the author

Table 7.7 Employment multiplier effects of studies relying on input-output analysis

Country	Study	Type I	Type II
EE.UU.	Crandall et al. (2003)	N.A.	2.17
	Atkinson et al. (2009)	N.A.	3.60
	Katz et al. (2009)	1.83	3.42
Switzerland	Katz et al. (2008)	1.38	N.A.
United Kingdom	Liebenau et al. (2009)	N.A.	2.76
Germany	Katz et al. (2009)	1.45	1.93

Note Crandall et al. (2003) and Atkinson et al. (2009) do not differentiate between indirect and induced effects, therefore we cannot calculate Type I multipliers; Katz et al. (2008) did not calculate Type II multiplier because induced effects were not estimated
Source Adapted from Katz (2009a)

These studies have provided the theoretical framework and tools to estimate the overall job impact of broadband-enabled counter-cyclical programs. For example, In 2010, Australia embarked on the construction of a National Broadband Network with a total funding estimated at AUD40.9 billion over 8 years to build and operate a new open access wholesale network. The plan will support up to 25,000 jobs over the life of the project. Funding will initially come from government, which will contribute equity of AUD27.5 billion, with other funding expected to come from operational earnings and private debt. In 2015, the wholesale network provider (NBN Co) will begin raising funds through capital markets, with an estimated AUD13.4 billion required to finance the project.

7.1.2 Broadband Spillovers on Business Expansion and Innovation

Beyond the employment impact of network construction, researchers have also studied the impact of network externalities on job creation variously categorized as "innovation," or "network effects."[2] Once deployed extensively, broadband becomes a "general purpose technology" with the power to facilitate growth and innovation throughout the economy. The study of network externalities resulting from broadband penetration has led to the identification of numerous effects:

- New and innovative applications and services, such as telemedicine, Internet search, e-commerce, online education, and social networking[3]
- New forms of commerce and financial intermediation[4]
- Mass customization of products[5]

---

[2] Atkinson et al. (2009).

[3] Op. cit.

[4] Op. cit.

[5] Op. cit.

- Marketing of excess inventories and optimization of supply chains[6]
- Business revenue growth[7]
- Growth in service industries[8]

The results of microeconomic research have been utilized to estimate the impact of broadband on job creation. For example, Fornefeld et al. (2008) identified three ways that broadband impacts employment: first, the introduction of new applications and services causes acceleration of innovation in terms of creation of new ventures and natural business expansion requiring additional workers; second, the adoption of more efficient business processes enabled by broadband increases productivity, which in term, tends to reduce employment; and third, the ability to process information and provide services remotely makes it possible to attract (or loose) employment from other regions through outsourcing. According to Fornefeld et al. (2008), these three effects act simultaneously, whereby the productivity effect and potential loss of jobs due to outsourcing are neutralized by the innovation effect and gain of outsourced jobs from other regions.

Thus, according to the authors, the negative effect of broadband productivity is compensated by the increase in the rate of innovation and services, thereby resulting in the creation of new jobs. The third effect may induce two countervailing trends. On the one hand, a region that increases its broadband penetration can attract employment displaced from other regions by leveraging the ability to relocate functions remotely. On the other hand, by increasing broadband penetration, the same region can lose jobs by virtue of the outsourcing effect. While we are gaining a better understanding of these combined "network effects," the research is still at its initial stages of quantifying the combined impact. The study by Fornefeld et al. (2008) is probably the first attempt to build a causality chain. It applies ratios derived from microeconomic research to estimate the combined impact of all effects, rather than conducting aggregate econometric analysis of historical data.

Most of the research regarding the impact of broadband externalities on employment has been conducted using US data. There are two types of studies of these effects: regression analyses and top-down multipliers. The first ones attempt to identify the macroeconomic variables that can impact employment,[9] while the second ones rely on top-down network effect multipliers.

Among the econometric studies of employment impact, are Gillett et al. (2006), Crandall et al. (2007), Shideler et al. (2007) and Thompson and Garbacz (2008). Relying on standard regression analysis, a team of MIT and Carnegie Mellon researchers[10] conducted a study that sought to measure the impact of broadband on

---

[6] Op. cit.

[7] Varian et al. (2002), Gillett et al. (2005).

[8] Crandall et al. (2007).

[9] In general, studies based on regression analysis do not differentiate between construction and spillover effects.

[10] Gillett et al. (2006).

a number of economic indicators including employment. Initially, the study tried to estimate the relationship between broadband and employment at the state-level. However, it concluded that data at this level of aggregation did not permit observation of any measurable impact. It was only when they turned to the zip code level that a positive impact of broadband was observed: the availability of residential broadband added over 1.5 % to the employment growth rate in a typical community.[11]

Crandall et al. (2007) relied on the same methodology to conduct a study focused on the effects of broadband on output and employment for the 48 US states. The conclusion of their multivariate regression analysis was that "for every 1 % point increase in broadband penetration in a state, employment is projected to increase by 0.2–0.3 % a year (...) (an increase of about 300,000 jobs, assuming the economy is not already at "full employment").

Shideler et al. (2007) analyzed the relationship between broadband saturation and employment growth in counties of the US state of Kentucky. The authors found that the coefficient on broadband saturation is positive and statistically significant on aggregate county employment in several industrial sectors (details below). It ranged from 0.14 to 5.32 % for each 1 % incremental broadband penetration.

Finally, Thompson and Garbacz (2008) employed a stochastic-frontier production function to measure the direct and indirect impact of broadband penetration on the GDP 48 US states. While they found that employment in certain industrial sectors tends to grow with broadband penetration, they also pointed that broadband deployment may cause a substitution effect between capital and labor.

To sum up, after examining the conclusions of the regression studies, the evidence regarding broadband employment externalities appears to be quite conclusive (see Table 7.8).

Again, the impact of broadband on employment creation appears to be positive. However, as the data indicates, the impact on employment growth varies widely, from 0.2 to 5.32 % for every increase in 1 % of penetration. There are several explanations for this variance. As Crandall indicated, the overestimation of employment creation in his study is due to employment and migratory trends, which existed at the time and biased the sample data. In the case of Gillett et al. (2006), researchers should be careful about analyzing local effects because zip codes are small enough areas that cross-zip code commuting might throw off estimates on the effect of broadband. For example, increased wages from broadband adoption in one zip code would probably raise rent levels in neighboring zip codes prompting some migration effects. Finally, the divergent effects among industry sectors explain the wide range of effects in the case of Shideler et al. (2007).

---

[11] Because of their approach, the researchers did not differentiate between job effects (network construction vs. utilization).

Table 7.8 Research results of broadband impact on employment in the United States

Authors—Institution	Data	Effect
Crandall et al. (2007)— Brookings Institution	48 states for the period 2003–2005	For every 1 % point increase in broadband penetration in a state, employment is projected to increase by 0.2–0.3 % per year "assuming the economy is not already at 'full employment'"
Thompson and Garbacz (2008)—Ohio University	46 states during the period 2001–2005	Positive employment generation effect varying by industry
Gillett et al. (2006)—MIT	Zip codes for the period 1999–2002	Broadband availability increases employment by 1.5 %
Shideler et al. (2007)— Connected Nation	Disaggregated county data for state of Kentucky for 2003–2004	An increase in broadband penetration of 1 % contributes to total employment growth ranging from 0.14 to 5.32 % depending on the industry
Katz et al. (2011)	Disaggregated county data for Kentucky, West Virginia and Ohio between 2004 and 2007	An increase in broadband penetration of 1 % contributes to total employment growth ranging from 0.12 to 0.84 % depending on the industry

Source compiled by the authors

Beyond regression studies, "network effect" multipliers have been used to assess the impact of broadband on job creation through a top-down approach. Within this group, key studies are Pociask (2002), Atkinson et al. (2009), and Liebenau et al. (2009). Pociask (2002) and Atkinson et al. (2009) studies relied on an estimated "network effect" multiplier, which is applied to the network construction employment estimates. For example, Pociask relied on two multiplier estimates (an IT multiplier of 1.5–2.0 attributed to a think tank and another multiplier of 6.7, attributed to Microsoft) and calculated an average of 4.1. Similarly, Atkinson et al. (2009) derived a multiplier of 1.17 from Crandall et al. (2003). Though the top-down approach allows estimation of the broadband impact, it does not have a strong theoretical basis. Network effects are not built on interrelationships between sectors. They refer to the impact of the technology on productivity, employment, and innovation by industrial sector (Figs. 7.1 and 7.2).

The methodological implications of these studies are that in order to properly measure the contribution of broadband to job creation, it is advisable to have datasets that include time series for employment level, broadband penetration, and related human capital statistics at a disaggregated level, such as counties, departments, or administrative district.[12]

---

[12] See examples in Katz (2010a).

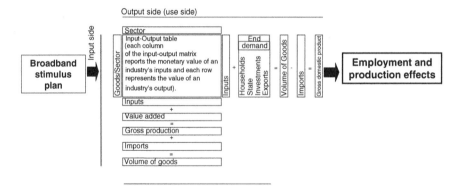

Fig. 7.1 Structure of Input-output table. *Source* Adapted from Katz (2012)

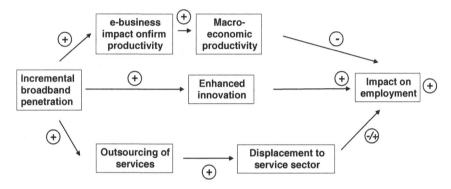

Fig. 7.2 Network effects of broadband on employment *Note* This causality chain was adapted from a model originally developed by Fornefeld et al. 2008 in a report for the European Commission

Like the relationship between broadband and GDP growth, the contribution of broadband to employment is also conditioned by a number of special effects. Studies have particularly focused on two specific questions:

- Does the impact on employment differ according to industry sector?
- Is there a decreasing return in employment generation linked to broadband penetration?

As with GDP, the spillover employment effects of broadband are not uniform across sectors.[13] According to Crandall et al. (2007), the job creation impact of broadband tends to be concentrated in service industries, (e.g., financial services, education, health care, etc.) although the authors also identified a positive effect in manufacturing. In another study, Shideler et al. (2007), supported by findings from Katz et al. (2011) found that, for the state of Kentucky, county employment was

---

[13] Differential employment effects by industry will be reviewed in detail in Sect. 7.3.

positively related to broadband adoption in the following sectors: health care, wholesale trade, and certain services. The only sector where a negative relationship was found with the deployment of broadband (0.34–39.68 %) was the accommodations and food services industry. This may result from a particularly strong capital/labor substitution process taking place, whereby productivity gains from broadband adoption yields reduced employment. Similarly, Thompson and Garbacz (2008) conclude that, for certain industries, "there may be a substitution effect between broadband and employment."[14] It should therefore be considered that the productivity impact of broadband can cause capital-labor substitution and may result in a net reduction in employment. However, the innovation effect contributing to new firm formation could still be at work.

Beyond aggregate estimates indicating positive externalities of broadband, disaggregated geographic effects are more complex. For example, a study by Katz et al. (2010) on the broadband externalities of Germany's National Broadband Strategy found that, while the positive effects on employment and economic output resulting from enhanced productivity, innovation, and value chain decomposition are significant throughout Germany; the job impact of broadband tends to vary over time and by region. By splitting the German territory into two groups, counties with 2008 average broadband penetration of 31 % of population and counties with average broadband penetration of 24.8 %, the analysis determined that the type and pattern of network effects of broadband varies by region. In high broadband penetrated counties the short-term impact of the technology is very high both on GDP and employment, but it declines over time. This "supply shock" is believed to occur because the economy can immediately utilize the new deployed technology. Furthermore, the fact that employment and GDP grow in parallel indicates that broadband has a significant impact on innovation and business growth, thereby overcoming any employment reduction resulting from capital/labor substitution effects.

On the other hand, in counties with low broadband penetration the impact of broadband on employment is slightly negative in the initial years. Negative initial employment growth appears to indicate that the productivity increase resulting from the introduction of new technology is the most important network effect to begin with. However, once the economy develops, the other network effects (innovation and value chain recomposition) start to play a more important role, resulting in job creation.[15] Therefore broadband deployment in low penetrated areas will likely generate high stable economic growth ("catch up" effect) combined capital/labor substitution, which initially limits employment growth ("productivity" effect). Figure 6.39 presents in conceptual fashion a comparison of impact in both regions.

The finding conceptualized in Fig. 7.3 is consistent with Lehr et al. (2005) and Thompson and Garbacz (2008) finding that there is a short-term negative impact of broadband on employment due to process optimization and capital-labor

---

[14] This effect was also mentioned by Gillett et al. (2006).

[15] This said, the available data sets do not enable us to test this last point at this time.

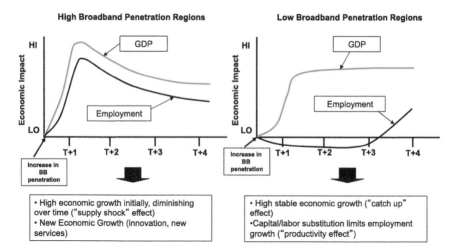

Fig. 7.3 Conceptual view of comparative broadband regional effects only effects up to t + 3 are estimated. *Source* Adapted from Katz et al. (2010)

substitution. However, the German evidence appears to point out that the short-term negative effect is limited to low broadband penetration geographies. These differentiated effects are critical in generating evidence regarding job creation of broadband in rural areas. The results of the research appear to be mixed in this regard.

In the first place, in response to the argument that broadband deployment in rural areas stimulates the relocation of establishments to those areas and therefore creates jobs, evidence indicates that relocation as a result of broadband tends to occur to the suburban peripheries of metropolitan areas rather than the strictly rural geographies (Katz et al. 2011). Furthermore, research also indicates that the jobs created in the periphery are jobs lost in metropolitan areas ("job cannibalization" effect), which means that no additional jobs are created. For example, Lehr et al. (2005) assessed a district-level (zip code) panel of data on broadband deployment in the United States and concluded that zip codes with broadband experienced faster job growth rate (1–1.4 %) between 1998 and 2002, experienced faster growth in business establishments (0.5–2.2 %), and a favorable shift in the mix of business toward higher value-added ICT intensive sectors (0.3–0.6 %). In response to this finding, some economists argue that, "job cannibalization" resulting from firm relocation triggered by broadband technology yields equilibrium in labor markets (and therefore has a positive contribution).

Second, research indicates that job effects of broadband in rural areas are highly dependent on industrial sectors. As we will show below, job creation effect in metropolitan peripheries is present in wholesale trade and financial services, while broadband-induced employment growth in rural areas is only significant in retail trade, and health services.

In sum, the employment impact of broadband by geography appears to indicate the existence of multiple effects at work (see Fig. 7.4).

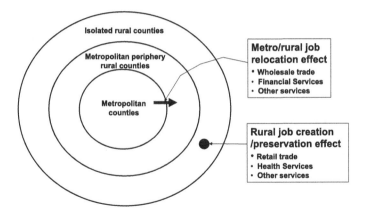

Fig. 7.4 Geographic impact of employment triggered by broadband deployment

Table 7.9 Brazil: Variables utilized to estimate the impact of broadband on job creation

Variable	Series	Source	Observations
Change of unemployment rate	Change in unemployment rate (2006–2007)	IBGE	Dependent variable
Control for level of development	GDP per capita (2003)	IBGE	Variable to determine the point of departure of state economic growth
Growth in household broadband penetration	Growth rate in broadband penetration (2005–2006)	Household survey (IBGE)	Independent variable
Control for human capital	Years of schooling	IBGE	Variable to differentiate the level of human capital by state
Control for population growth	Population growth (2006–2007)	IBGE	Variable to differentiate the level of population growth by state

Source Adapted from Katz (2010a)

7.1.3 Broadband and Employment in Developing Countries

Beyond the studies conducted in mature economies, the aggregate impact of broadband on employment has begun to be studied in emerging countries as well. For example, Katz (2012) constructed a cross-sectional sample for the 27 Brazilian states similar to that one utilized by Katz et al. (2010) for Germany, where the dependent variable was the rate of change of unemployment:

The results of the model are as follows (see Tables 7.9 and 7.10).

However, the relationship between the rate of change of unemployment rate and the rate of change in broadband penetration is significant and with the expected

Table 7.10 Brazil: impact of broadband on job creation

Unemployment rate	Coefficient	Standard error	T-statistic	$P > [t]$
Control for level of state economic development	−0.0449	0.0259	−1.73	0.098
Growth in household broadband penetration	−0.0069	0.0036	−1.94	0.066
Control for human capital	0.1095	0.0940	1.17	0.256
Control for population growth	0.2009	0.1213	1.66	0.112
Constant	−0.1925	0.5035	−0.38	0.706
Number of observations	27			
F(4, 22)	3.76			
Prob > F	0.0178			
R^2	0.4058			
Root MSE	0.27016			

Source Adapted from Katz (2010b)

negative sign. According to the model results, a change of 10 % in broadband penetration could reduce the unemployment rate by 0.06 % points. On the other hand, it is counterintuitive that the difference in schooling years and population growth are not statistically significant in explaining the differences in the level of unemployment across regions.

The availability of an extensive database of quarterly data for Chile allowed the development of a panel of time series data for each of Chile's administrative regions. This database was constructed by compiling data for each of Chile's regions (except for the Metropolitan Region due to the lack of quarterly data) from 2001 until the fourth quarter of 2009. The dataset contains the following information (Table 7.11):

A model including level of economic activity and broadband penetration was specified. In addition, an alternative model was proposed aimed to study possible effects of human capital and specialization on the level of employment. According to the methodology used, the fixed effects of the panel data control other specific characteristics of each region that could have an impact on the labor market. Thus, the model results are as follows (see Table 7.12).

The economic activity variable remained unaltered between the two specifications; in both cases it was significant and positive. The coefficient of broadband penetration is significant[16] and positive in both specifications. The small variation between the two specifications suggests that the contribution of 1 % point in broadband penetration would contribute nearly 0.18 % points to the employment rate.

An interesting result on the second specification was the coefficient of human capital, which resulted significant at the 10 % level and with a negative sign.

---

[16] It was significant at the 1 % level on the first specification and at the 5 % level on the second one.

Table 7.11 Chile: Variables used to estimate the broadband impact on job creation

Variable	Series	Source	Observations
Employment	Quarterly employment rate (2002–2009)	Regional institutes of statistics	Dependent variable
Control for level of economic activity by region	Quarterly Index of economic activity (2001–2009)	Regional institutes of statistics	
Growth in broadband penetration	Quarterly growth in broadband penetration (2002–2009)	Subtel	Independent variable
Human capital	Schooling Years (population 15 years old and older)	Employment survey, INE	Independent variable
Dominant sectors	Contribution of the mining and financial sector to regional GDP (2002–2008)	Central bank of Chile	Variable to control for regional specialization in dominant economic activities
Dynamic sectors	Contribution of the agricultural and trade sector to regional GDP (2002–2008)	Central bank of Chile	Variable to control for regional specialization in dynamic economic activities

Source Katz (2010a)

Table 7.12 Chile: Broadband impact on job creation

Employment rate	Model 1		Model 2	
	Coefficient	t-statistic	Coefficient	t-statistic
Level of economic activity by region	0.0003	5.90	0.0003	5.72
Growth in broadband penetration	0.1812	3.85	0.1774	2.56
Human capital			−0.0042	−1.87
Dominant sectors			−0.0013	−1.66
Dynamic sectors			0.0017	1.27
Constant	0.8682	109.03	0.9138	25.95
Model 1 number of observations	324	Model 2 number of observations	276	
$F_{(2, 310)}$	60.89	$F_{(5, 259)}$	20.78	
Prob > F	0.0000	Prob > F	0.0000	
R2	0.2820	R2	0.2863	
$F_{(11, 310)}$	33.89	$F_{(11, 259)}$	24.41	
Prob > F	0.0000	Prob > F	0.0000	

Source Katz (2010)

Contreras and Plaza (2008) argues that this result is mostly explained by the impressive increment in the years of schooling of the population in one generation and the entrance of women to the labor force.[17]

A model similar to one constructed for Germany was developed by Katz et al. (2011) for Colombia. In this case, growth of the employment rate was studied in relation to the increase in broadband access lines, controlling for population growth and economic development. The model was specified both for the whole country and for departments with high and low broadband penetration.

In these models, the broadband contribution to job creation is significant at the national level and for the low penetrated departments; in the highly penetrated departments, statistical significance is only 24 %. On the other hand, population growth appears to have an effect only in high penetration departments (with a positive coefficient). This could be explained by the fact that in highly penetrated departments it is easier to be placed in the labor market, according to the innovation effect reviewed above (Table 7.13).

Finally, Katz (2012) conducted a study on job creation in the Dominican Republic, relying on panel data for the country's 32 provinces. Contrary to the Chilean and Colombian models, the objective in this case was to study broadband contribution to the reduction of unemployment. The results in this case exhibit high impact. An increase in broadband penetration of 1 % reduces unemployment in 0.29 % points. The other variables affecting unemployment in an indirect fashion are, as expected, the change in the number of industrial establishments, and the growing importance of the construction sector.

According to Table 7.14, an increase in broadband penetration of 1 % would diminish unemployment in 0.29 % points. For example, if the unemployment rate were to be 14 %, an increase of 1 % in broadband penetration would contribute to a reduction of unemployment to 13.71 %.

The other variables that indirectly affect unemployment are, as expected, number of establishments between 2008 and 2009, and the intensity of the construction sector in a specific area in 2009. Therefore, a combination of increase in the number of establishment, investment in construction and broadband yields a positive effect in terms of job creation.

It is considered that the contribution of broadband relative to the other two variables is too high. Part of this is due to the fact that the largest increase in broadband has taken place in Santo Domingo, the capital, and Altagracia, a tourism hub. Ideally, to refine the impact of broadband it would be necessary to include in the model specification a variable related to the intensity of the tourism sector. However, while data exists for three regions, Altagracia, Puerto Plata, and

---

[17] According to Contreras and Plaza (2008), women were educated but were not part of the labor force. As more women entered the labor force the average schooling years of the population increased but also increased the number of women unemployed looking for jobs.

Table 7.13 Colombia: Impact of increase in broadband penetration on job creation

Growth of employment rate (%)

Dependent variable: growth of employment rate (2006–2010)

Independent variables: growth in broadband connections (2006–2010), population growth (2006–2010), GDP 2003

	Total	Low penetration	High penetration
Growth in broadband connections (%)	0.0003*	0.0003**	0.0006572
	(0.0001359)	(0.0001547)	(0.0005495)
Population growth (2006–2010) (%)	0.0160	−0.2539	0.5937073***
	(0.5114836)	(0.7899623)	(0.3761862)
GDP 2003 (in million pesos)	0.0053	−0.1085	0.0003309
	(0.0077051)	(0.1308956)	(0.0090124)
R^2 adjusted	0.0110	0.0318	0.0338
Prob > F	0.0730	0.0321	0.4351
Number of observations	132	64	68

Note Low penetration departments refers to administrative districts whose broadband penetration is lower than the country mean, while high penetration refers to departments whose broadband penetration is higher than the country mean
Source Katz y Callorda (2011). Symbols *, ** and *** indicate statistical significance at 15, 10 and 5 % respectively

Table 7.14 Dominican Republic: Impact of Growth in Broadband penetration on unemployment reduction

Unemployment growth	Coefficient	Standard error	T-statistic	P > t	95 % confidence interval	
Population growth	0.7244	0.24939	2.90	0.0070	0.21180	1.23704
Change in broadband penetration	−0.2953	0.13290	−2.22	0.0350	−0.56846	−0.02211
Change in number of establishments	−0.1496	0.04728	−3.16	0.0040	−0.24678	−0.05241
Value of construction sector 2009	0.6946	0.14588	4.76	0.0000	0.39469	0.99443
Change in value of construction 2008–2009	−0.6430	0.12787	−5.03	0.0000	−0.90583	−0.38015
Constant	0.7432	0.37360	1.99	0.0570	−0.02477	1.51111
Number of observations	32					
F(5, 26)	12.70					
Prob > F	0.0000					
R^2	0.4175					

Source Katz (2012)

Cibao Norte, data for the rest of the country is not captured. Therefore, it is impossible to introduce a variable measuring the intensity of tourism. As a result, the case confirms the contribution of broadband to job creation, although the range of impact might be overestimated.

Table 7.15 Comparative estimate of broadband impact on employment growth

Study	Region/ Country	Impact on job creation for each incremental 1 % in rate of growth of broadband penetration		Observations
		Employment	Unemployment	
Katz (2012)	Brazil		−0.00691	• Statistically significant coefficient (t-statistic = −1.94) • 27 observations
Katz (2012a)	Chile	0.00181		• Statistically significant coefficient (t-statistic = 3.85) • 324 observations
Katz and Callorda (2011)	Colombia	0.00030		• Satistically significant coefficient • 132 observations
Katz (2012)	Dominican Republic		−0.29529	• Statistically significant coefficient (t-statistic = −2.22) • 32 observations

Source Compiled by the authors

The results of the analyses also lead to validate the positive contribution of broadband to employment creation for less developed countries and regions. The following chart compiles the results of the existing literature (see Table 7.15).

In this case, all prior research and as well as the results of this study coincide that broadband has a positive impact on job creation. In particular, the Chilean and the Colombian cases, which are based on an extensive datasets, yield statistically significant positive coefficients. The other cases (Brazil and Dominican Republic) have also yielded statistically significant coefficients for the explanatory variable (broadband penetration) with sensible signs—positive when the independent variable is employment and negative when it is unemployment. While these studies are country-specific and cannot be applied directly to other nations, they provide an estimate of the potential employment gains that could result from effective broadband development. The positive relationship between broadband development and job growth is not in question. However, the magnitude of the impact should be the subject of further analysis, as the present studies are compiled using different methodologies and data samples.

Coincident with these empirical studies, governments in the emerging world have begun to estimate the employment impact of broadband development programs. For example, the Malaysian Communications and Multimedia Commission (MCMC) estimated in 2008 that achieving 50 % broadband penetration by 2010 could create 135,000 new jobs in the country. The MCMC further projected that the number of jobs created would reach 329,000 by 2022, based on 50 % broadband penetration rate.

7.2 Broadband, Productivity, and Firm Efficiency

Converging with the aggregate macroeconomic research, the microeconomic analysis of the impact of broadband has helped understand the multiple effects that the technology has on firm performance. For example, several studies have been conducted to determine the relationship between broadband and productivity. Atrostic and Nguyen (2006) analyzed the productivity of 25,000 manufacturing plants according to data compiled by the US Census Bureau. They found a correlation in the range of 3.85–6.07 % between intense use of business processes enabled by the introduction of broadband and labor productivity. Rincón-Aznar et al. (2006) applied a similar methodology. The authors relied on a database of enterprises using e-Business, which was compiled by the UK National Institute of Economics and Social Research. They found that the average impact of such processes (enabled by broadband access) resulted in 90 % of firms in the service sector improving their productivity by 9.8 %. The researchers also found that productivity improvement tends to be higher in certain segments of the service sector.

These results are consistent with the analysis of the e-Business watch survey, which determined that e-Business productivity gains depend on the industrial sector. The variable found to better explain the amount of broadband impact is the level of information-intensity of business functions. Based on this conclusion, Fornefeld et al. (2008) in a study commissioned by the European Commission estimate that broadband increases productivity of information-intensive firms by 20 %. To summarize, microeconomic research has yielded the following estimates of firm productivity enhancement (see Table 7.16).

In addition to the impact on productivity, other microeconomic studies have focused on the impact of broadband technology on business expansion, product innovation, and new business creation. With regard to business expansion, Clarke (2008) studied the impact of broadband access on exports of manufacturing and service firms. The analysis was performed for countries of medium and low levels of development in Eastern Europe and central Asia. The study controlled for variables such as firm size, industrial sector, foreign ownership, firm performance, level of domestic competition, international trade organization affiliation, progress in privatization, and telecommunications infrastructure. The author found that in the manufacturing sector firms with Internet access enabled by broadband generate 6 % more foreign sales than the rest. In the service sector, broadband enabled firms generate between 7.5 and 10 % more sales.

In addition to increasing exports, broadband has been found to have a positive impact on the development of new businesses. This results from the network effects of connectivity. When a large enough number of households are connected to broadband, the incentive to develop new businesses around information search, advertising and electronic commerce increases. For example, Crandall et al. (2007)

Table 7.16 Broadband-induced Productivity Improvement

Industrial sector	Study	E business impact on firm productivity (%)	Share of informational activities that involve external parties (%)
Manufacturing	Atrostic and Nguyen (2006)	~5	~25
Services	Rincón-Aznar et al. (2006)	~10	~50
Information	Fornefeld et al. (2008)	~20	100

Source Fornefeld et al. (2008)

estimate that the network effects of universal broadband access can have a multiplier of 1.17 on the investment in infrastructure. As a result of 40 % lower broadband penetration in the United Kingdom, Liebenau et al. (2009) estimate the multiplier to be somewhat lower (0.33) for the British economy.

7.3 Industrial Sectors Most Impacted by Broadband

As with output, the spillover employment effects of broadband are not uniform across sectors. Two studies have identified differential levels of impact. According to Crandall et al. (2007), the job creation impact of broadband tends to be concentrated in service industries, (e.g., financial services, education, health care, etc.) although the authors also identified a positive effect in manufacturing (see Table 7.17).

In another study, Shideler et al. (2007) found that, for the state of Kentucky, county employment was positively related to broadband adoption in the following sectors (see Table 7.18).

The only sector where a negative relationship was found with the deployment of broadband (0.34–39.68 %) was the accommodations and food services industry. This may result from a particularly strong capital/labor substitution process taking place, whereby productivity gains from broadband adoption yields reduced employment. Crandall et al. (2007) also found a negative relationship for the Arts, Entertainment and Recreation sector, although it was not statistically significant.

Similarly, Thompson and Garbacz (2008) concluded that, for certain industries, "there may be a substitution effect between broadband and employment."[18] It should therefore be considered that the productivity impact of broadband can cause capital-labor substitution and may result in a net reduction in employment.

In summary, research is starting to pinpoint different employment effects by industry sector. Broadband may simultaneously cause labor creation triggered by innovation in services and a productivity effect in labor-intensive sectors.

---

[18] This effect was also mentioned by Gillett et al. (2006).

Table 7.17 United States. coefficient of broadband penetration in employment growth by sector (with significance at the 5–1 % confidence level)

Sector	Employment 2005–2014		Employment 2005–2013	
	Coefficient	t-statistic	Coefficient	t-statistic
Manufacturing	0.371	2.46	0.789	2.59
Educational services	2.741	2.73	4.054	3.25
Health care	3.369	2.50	0.656	2.51
Accommodation and food services	0.284	2.12	N.A.	
Finance and insurance	N.A.		1.043	3.09

N.A. Statistically not significant
Source Crandall et al. (2007)

Table 7.18 Kentucky: differential impact of broadband by industry sector

Sector	95 % confidence interval (%)
Aggregate	0.14 5–5.32
Construction	0.62–21.76
Information	25.27–87.07
Administrative	23.74–84.56

Source Shideler et al. (2007)

However, we still lack a robust explanation of the precise effects by sector and the specific drivers in each case. However, given that the sectoral composition varies by regional economies, the deployment of broadband should not have a uniform impact across a national territory.

7.4 Broadband and Enterprise Relocation

Enterprise geographic strategies are determined by several objectives, not all consistent. At the most basic level, firms define their location as a function of a market they want to serve. Closeness to the target market allows better understanding of customer needs and faster responsiveness to environmental changes. At the same time, firms need to consider in their location strategy an optimal approach to accessing a valuable resource (labor, raw materials). Closeness to the supply input builds competitive advantage (also called static arbitraging). Finally, firms tend to cluster in certain geography as a way of lowering transaction costs. While counterintuitive (since competitors in a common cluster might fight for inputs and talent), the need to trade among themselves for inputs, and attracting the investment of suppliers makes it highly convenient to collocate (Figs. 7.5 and 7.6).

Meeting all three locational requirements might not be easy. If a firm prioritizes the lowering of transaction costs, that might put it far away from its customers. Alternatively, closeness to the market might run against static arbitraging (gaining access to valuable inputs).

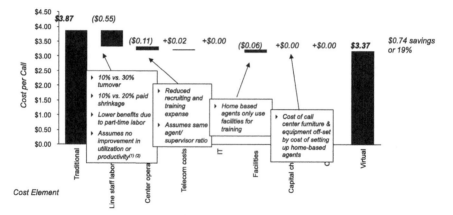

Fig. 7.5 Traditional versus Virtual Contact Center Cost Structure, Cost per Call Flexible staffing model can allow for 15 % + reduction in agent FTE over model using 8 h shifts Interviews indicate improvement in productivity from home-based agents *Note* based on financial services case with 80 % service level, 45 s ASA, 272 AHT, 3 million rep calls per year. *Source* Booz Allen Hamilton (2004)

Fig. 7.6 Number of Customer Service, Technical Support and Telemarketing Agents in the US and India, 2000–2008. *Source* US Department of Labor; Nasscom (India)

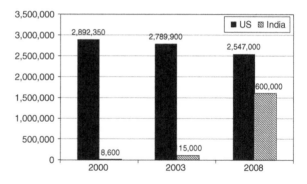

In this context, broadband technology represents a powerful tool to help defining a geographic strategy that maximizes all three conditions. First, broadband reduces the communication costs that buyers and sellers incur to complete a transaction. In that sense, closeness to the market might not be such a strong imperative. Second, broadband can reduce the search costs for a particular input, rendering static arbitraging not as advantageous at least for all inputs. At the same time, broadband allows companies to relocate to places that are attractive, with the implication that good quality of life helps attracting talent. In sum, broadband allows firms to define their locational strategy by prioritizing one of the three geographic constraints without a concern of losing to its competitors. There are several examples of enterprises that succeed despite the fact they are not collocated within a given industrial cluster: Microsoft in Washington State, not in

Silicon Valley, Berkshire Hathaway in Omaha, Nebraska, rather than Wall Street are two examples.

In addition to enabling a decision to move geographically, broadband fulfills another important function. When considering a company's value chain, broadband enables the firm to relocate certain functions to meet some locational requirements without necessary increasing its coordination costs. Along those lines, manufacturing could be located close to the source of inputs to reduce cost and optimize the supply chain, while sales and marketing are located close to the markets being served. Before the Internet, these geographic strategies could be the source of increasing complexity, such as problems in the interface between marketing, product design, and manufacturing. This was one of the key advantages Matsushita had over Phillips by collocating all its primary functions in Osaka, Japan. Broadband-enabled systems allow now to geographically fragmenting the firm's value chain without incurring higher complexity costs. Examples of this approach range from companies like Nike footwear (manufacturing in Asia while design is located in the United States) to the Indian systems integrators (serving the European and North American markets with talent from India).

7.4.1 Broadband and Entrepreneurship

If, as pointed above, broadband is causally linked to innovation, it would stand to reason that the technology would have an impact on entrepreneurship, defined as the incentive to launch a new business. The causality would operate at two levels. On the one hand, broadband adoption opens new market reach opportunities to the development of new service offerings (such as Internet based businesses). As such, Venkataraman (2004) argues that broadband infrastructure is a prerequisite for high tech venture growth. In addition, broadband remains a necessary infrastructure supporting development teams that are at the center of new business incubation. For example, Audretsch (1998) argues that information flows necessary to incubate new businesses are accelerated by broadband infrastructure. Finally, broadband can generate savings in start-up costs for the entrepreneur. According to the Internet Innovation Alliance (IIA) and the Small Business and Entrepreneurship Council, start-up savings generated by broadband could amount to over $16,000, comprising accounting services, printing, telecommunications, web design and hosting, and other applications (Kerrigan and Keating 2012). Along these three lines of reasoning, it would appear that broadband deployment would foster innovation and entrepreneurship, with a positive contribution on economic growth.

The study of the relationship linking broadband and entrepreneurship has begun only recently. The evidence generated so far is relatively sparse, particularly in light of considerable data challenges. For example, in a study cited above, Gillett et al. (2006) found that broadband access tends to reduce the share of small business establishments by 1.3–1.6 %, which would imply a negative correlation.

However, the study authors also acknowledged that given the limited availability of data, "methodological challenges inherent in disentangling causality" remain. Similarly, Heger et al. (2011) used county-level data for Germany and found that broadband infrastructure does not have an impact on the overall level of entrepreneurial activities after controlling for regional characteristics. The only exception to this general finding was in high-tech industries (software and technology-intensive sectors), where higher broadband adoption was correlated with new business creation. A likely explanation of this trend is that the high-technology sector is quite dependent on efficient transfer of knowledge, which is naturally facilitated by broadband networks.

In a recent study, Carlsen and Zhou (2012) analyzed OECD data through a set of econometric techniques aimed at controlling for reverse causality and found that "a 1 % point increase in broadband penetration, ceteris paribus, results in an additional 0.086 new business registered per 1000 inhabitants of working age, a proxy for entrepreneurship. Such an increase in broadband penetration is in line with the average percentage point increase between 2008 and 2009 for the 23 OECD countries. For the average OECD country in our sample, this corresponds to 1,625 new firms. In Sweden, the estimated number of new firms is equivalent to 523, an increase of 2.2 %."

7.4.2 Broadband as an Impact on Employment Structure and Distribution on a Global Scale

In addition to its impact on job creation, broadband also affects the global job distribution and the structure of employment. As broadband is facilitating the globalization of service provisions in highly value-added services such as accounting and IT related services as well as lower skill services such as back-office functions and call center activities, jobs are migrating from high cost to low cost countries in these service sectors. In fact, the role of broadband in the international restructuring of job market in the service sector is increasing as more and more services can be traded as bandwidth increases. However, as Fornefeld et al. (2008) point out broadband-enabled outsourcing can benefit emerging regions benefitting from low cost labor and mature economies that leverage high-skilled workers.

For example, industrialized countries, such as the United States have been developing virtual call centers, which rely on home-based agents to complement/replace traditional call centers from emerging markets. This concept, enabled by home-based information technology and broadband infrastructure, has a number of distinctive advantages. It avoids overhead related to traditional contact centers, including real estate and utilities, as well as up-front investment/capital expenditures—particularly for rapid-growth companies. It lowers staff turnover due to work from home and flexible schedules, and improves service level and/or labor utilization due to flexible staffing—management of time of day and seasonal peaks. Finally, it provides greater responsiveness to unforeseen changes in demand and lowers labor factor cost by being able to attract talent at lower wages and benefits.

From an economic standpoint, a virtual call center located in an industrialized country offers 15–30 % cost advantage over the traditional centralized model (see Fig. 6.41).

With this cost differential, in addition to the capability of staffing native English speakers, the virtual model represents a broadband approach to reverse the emerging market outsourcing trend (see Fig. 6.42).

In addition to the outsourcing trend, the growth of broadband has also affected the internal employment structure of both, mature and emerging economies through the increase of self-employment by creating environment for individuals to deliver their services directly to customers through broadband. For example, according to the Small and Medium Business Administration (SMBA) of South Korea, the number of self-employed citizens in the nation reached 235,000 in 2010, upto 15.7 % from the preceding year, and the number now accounts for about 1 % of the economically active population. In the United States, 15.3 million workers (or 10.9 % of the economically active population) are self-employed. Of these, 28.2 % are in services such as information, financial activities, professional and business and education and health, all professions dependent to a large degree on broadband technology.

In addition to self-employment, broadband also assists with working from home. According to the Global Workplace Analytics, regular telecommuting in the United States grew by 73 % between 2005 and 2011 (3.1 million) compared to only 4.3 % growth of the overall workforce (not including the self-employed). In addition to the multiple benefits ranging from reduced carbon emissions to a reduction in real estate costs, telecommuting opens up the labor market to those who might be disadvantaged, for instance by disability or by living in a rural location. Broadband also enables many more part-time opportunities, for instance trading on eBay as a form of supplementary income, but not a full-time job. In fact, according to the July 2005 e-Bay survey, more than 724,000 Americans reported that eBay was their primary or secondary source of income. In addition to these professional eBay sellers, another 1.5 million individuals said they supplement their income by selling on eBay,

Another example of the impact of broadband on the job market is the emergence of the "virtual economy," revenue generating opportunities, conducted in an informal, decentralized setting, in new emerging areas like microwork, "gold farming" and gaming. The virtual economy is growing fast and contributes new job creations. As reported by Minn and Rossotto (2012), every year, about ten million people, half in the US, and the other half mainly in South Asia, earn an income from the microwork platform Amazon Turk, launched over 15 years ago. Microwork enables the division of a task into multiple "micro-tasks" and the microwork platform enables a client to ask a crowd of "micro-workers" to respond to the task. The Non-Governmental Organization (NGO) Samasource has extended the concept of microwork to highly distressed countries, including Southern Sudan. According to Samasource, each micro-worker in Sudan has been able to earn as much as $1,000 a year doing microwork, with a quality similar to the Amazon Turk worker.

7.5 Potential Negative Impact of Broadband on Employment and Management Strategies

As mentioned throughout this chapter, some researchers have found situations where broadband has a negative and/or decreasing impact of broadband on employment. In addition to job losses in some rural areas, Gillett et al. (2005) observed that, while the magnitude of impact of broadband on employment increases over time, they also found that the positive impact of broadband on employment tends to diminish as penetration increases. This finding may support the existence of a saturation effect. Coincidentally, Shideler et al. (2007) also found a negative statistically significant relationship between broadband saturation and employment generation. This indicates diminishing marginal returns.

In this context, it is useful to consider a set of additional policies to be put in place at the same time a broadband deployment program is executed:

1. Coordinate broadband deployment with job creation and retention programs: Network effects resulting from broadband programs can be sizable. However, their fulfillment is driven by success in implementing job creation and retention programs in parallel with network deployment. As an example, State and Local Governments in the targeted areas need to work with private sector companies in using this new infrastructure for employment generation. Also governments need to work with businesses to discourage job relocation as a result of broadband deployment.

   In addition, it is critical to deploy initiatives aimed at the creation of jobs enabled by broadband technology. Following on the example mentioned above, governments should stimulate the development of rural virtual call centers as a way to bring jobs that were outsourced overseas.

2. Rethink criteria for selecting areas to develop broadband: Consider deployment not only on unserved and underserved areas but also in regions where the possibility of developing regional growth, in coordination with broadband deployment, could act as a magnet to stimulate relocation, firm creation, and, consequently, jobs. While it is possible that private operators have already targeted such areas, it is reasonable to consider that opportunities for regional core development could be found. The experience of Germany, Sweden, and the Netherlands could be very instructive in this regard.

3. Centralize program evaluation and grant allocation: As a corollary to the first recommendation, given that the ability to generate jobs as a result of network externalities is dependent on the regions being targeted, it would be advisable

to centralize the process of allocating funds for network deployment and rely on a common framework for evaluating requests focused on economic growth and job creation. In this context, it is critical to enhance the government's ability to monitor spending and results, especially if the stimulus program is largely mandated like an earmark as opposed to some other methods that have more controls.

4. Develop a systematic test based on social and economic criteria to evaluate the return of the investment: All submissions for grants/loans should be backed up with analysis of the social and economic returns supported by a common set of tools and benchmarks.

5. Evaluate the economic impact of NGAN: This study has not quantified the effect of faster access speeds resulting from FTTx and/or DOCSIS 3.0. Given that no research has been conducted to date in this area, it is important to launch some analysis in this area.

References

Atkinson, R., Castro, D., Ezell, S.: The digital road to recovery: a stimulus plan to create jobs, boost productivity and revitalize America. http://archive.itif.org/index.php?id=212 (2009). Accessed 23 Mar 2014

Atrostic, B.K., Nguyen, S.: Computer input, computer networks, and productivity. In: Berndt, E.R. Hulten, C.R. (eds.) (2007). Hard-to-Measure Goods and Services: Essays in Honor of Zvi Griliches. University of Chicago Press, Chicago (2006)

Audretsch, B.: Agglomeration and the location of innovative activity. Oxford Rev. Econ. Policy **14**(2), 18–29 (1998)

Booz Allen Hamilton: Opportunity Assessment for Iowa Virtual Contact Center. Unpublished Report (2004)

Carlsen, L., Zhou, K.: Broadband and Entrepreneurship A cross-country study of 23 OECD countries between 2004 and 2009. Stockholm School of Economics, Stockholm (2012)

Contreras, D., Plaza, G.: Female Labor Force Participation in Chile: How Important Are Cultural Factors? (2008)

Crandall, R., Jackson, C., Singer, H.: The effect of ubiquitous broadband adoption on investment, jobs, and the US economy. Criterion Economics LLC (2003)

Crandall, R., Lehr, W., Litan, R.: The effects of broadband deployment on output and employment: a cross-sectional analysis of U.S. Data. http://www.brookings.edu/research/papers/2007/06/labor-crandall (2007). Accessed 23 Mar 2014

Clarke, G.: Has the internet increased exports for firms from low and middle-income countries? Inf. Econ. Policy, **20**(1), 16–37 (2008)

Fornefeld, M., Delaunay, G., Elixmann, D.: The impact of broadband on growth and productivity. A study on behalf of the European Commission (2008)

Gillett, S., Lehr, W., Osorio, C., Sirbu, M.: Measuring Broadband's Economic Impact. [report] United States Department of Commerce (2006)

Heger, D., Rinawi, M., Veith, T.: The Effect of Broadband Infrastructure on Entrepreneurial Activities. ZEW, Zentrum für Europäische Wirtschaftsforschung, Mannheim (2011)

Katz, R.L.: Estimating broadband demand and its economic impact in Latin America. Paper presented at ACORN-REDECOM Conference in Mexico City, 4–5 Sept 2009 (2009a)

Katz, R., Suter, S.: Estimating the economic impact of the broadband stimulus plan. Columbia Institute for Tele-Information Working Paper (2009)

Katz, R.L.: La contribucion de la banda ancha al desarrollo economico. In: Jordan, V., Galperin, H.Y., Peres, W. (eds.) Acelerando la revolucion digital: banda ancha para America Latina y el Caribe. Santiago: Cepal (2010a)

Katz, R.L.: The impact of broadband policy on the economy. Paper presented at the IV Conference of ACORN-REDECOM in Brasilia, 14–15 May 2010 (2010b)

Katz, R.L.: The impact of broadband on the economy: research to date and policy issues. Geneva: International Telecommunication Union (2012)

Katz, R.L., Callorda, F.: Medición de Impacto del Plan Vive Digital en Colombia y de la Masificación de Internet en la Estrategia de Gobierno en Línea. Bogotá (2011)

Katz, R.L., Vaterlaus, S., Zenhausern, P., Suter, S.: The impact of broadband on jobs and the German economy. Intereconomics **45**(1), 26–34 (2010)

Katz, R.L., Zenhausern, P., Suter, S.: An evaluation of socio-economic impact of a fiber network in Switzerland. [report] (2008)

Katz, R.L., Avila, J., Meille, G.: Economic impact of wireless broadband in rural America. [report]. Rural Cellular Association. Washington, D.C (2011)

Kerrigan, K., Keating, R.: Start-Up savings: boosting entrepreneurship through broadband internet. [report] Internet Innovation Alliance and Small Business & Entrepreneurship Council (2012)

Lehr, W., Osorio, C., Gillett, S.: Measuring broadband's economic impact. Paper presented at 33rd research conference on communication, information, and internet policy (TPRC), Arlington, 23–25 Sept 2005

Liebenau, J., Atkinson, R.D., Kärrberg, P., Castro, D., Ezell, S.J.: The UK's digital road to recovery. http://papers.ssrn.com/sol3/papers.cfm?abstract_id=1396687 (2009). Accessed 23 Mar 2014

Min, W., Rossotto, C.M.: Broadband and job creation: Policies promoting broadband deployment and use will enable sustainable ICT-based job creation. ICT Policy Notes: No 1. [report] The World Bank (2012)

Pociask, S.: Building a nationwide broadband network: speeding job growth. http://lus.org/uploads/BuildingaNationwideBroadbandNetwork.pdf (2002) Accessed 23 Mar 2014

Qiang, C.Z.: Broadband infrastructure investment in stimulus packages: relevance for developing countries. Info **12**(2), 41–56 (2010)

Rincon-Aznar, A., Robinson, C., Vecchi, M.: The productivity impact of e-commerce in the UK: 2001 evidence from microdata. NIESR (2006)

Shideler, D., Badasyan, N., Taylor, L.: The economic impact of broadband deployment in Kentucky. Paper presented at telecommunication policy research conference, Washington D.C (2007)

Thompson, H. Garbacz, C.: Broadband Impacts on State GDP: Direct and Indirect Impacts. Paper presented at international telecommunications society 17th biennial conference, Montreal (2008)

Venkataraman, S.: Regional transformation through technological entrepreneurship. J. Bus. Ventur. **19**(1), 153–167 (2004)

Varian, H., Litan, R.E., Elder, A., Shutter, J.: The net impact study. Cisco Systems Inc, San Jose (2002)

Chapter 8
Strategies to Promote Broadband Demand

The purpose of this chapter is to provide guidelines on how to introduce demand stimulation targets and policies in national broadband plans and national ICT planning in general. Some of the terms presented in this chapter are closely linked to concepts introduced in modules 1—Policy Approaches to promoting Broadband Development and 3—Law and Regulation for a Broadband World.

8.1 Determining Adoption Targets in National Broadband Plans

Reflecting the shift in emphasis from supply to demand, many national broadband plans now prioritize demand stimulation over infrastructure deployment. In general, this shift is more prevalent in developed nations, where the broadband coverage challenge has been addressed to a large degree.

The differences in targets can, in many cases, be related to the lack of rigor in setting them up. It is critical that countries follow a consistent and structured approach to defining targets.

The first area to be addressed regarding demand stimulation in national broadband plans pertains to the stipulation of adoption targets, which should complement the supply side (e.g., coverage of service) targets. These will have to be categorized in terms of residential, social (educational and health delivery institutions), enterprises (particularly SMEs), and public administrations. Furthermore, targets will have to be specified for both fixed and mobile broadband (Table 8.1).

In order to highlight broadband demand stimulation targets, it is advisable to structure goals on the basis of the following matrix (see Table 8.2).

Supply targets The first step in setting up broadband targets is to define the network coverage objective. This should be conducted for wireline and wireless networks. To a large degree, the social and residential coverage objectives have an impact on those of enterprises and the public administration sector, since

R. L. Katz and T. A. Berry, *Driving Demand for Broadband Networks and Services,*   295
Signals and Communication Technology, DOI: 10.1007/978-3-319-07197-8_8,
© Springer International Publishing Switzerland 2014

Table 8.1 National Broadband targets

Country	Targets
Australia	By 2021, the National Broadband Network will cover 100 % of premises, 93 % of homes, schools and businesses at up to 100 Mbps over fiber, with the remainder at up to 12 Mbps over next generation wireless and satellite
Austria	By 2013, 100 % of population will be provided with access speeds of at least 25 Mbps
Belgium	By 2015, 90 % of families to have broadband and 50 % of residents to be using the mobile Internet
Brazil	By 2014, to have 30 million fixed broadband connections, including homes, businesses and co-operatives, plus 100,000 telecenters
Chile	By 2011, to provide Internet access to 3 million rural households. By 2014, 100 % of school and 70 % of households to have broadband. By 2018, 100 % of households
China	By 2014, to raise broadband accessibility to 45 % of the population
Colombia	Reach 18 % of population adoption of broadband by 2014 with an offer of at least of 1 Mbps
Costa Rica	By 2014, achieve 100 % population coverage and by 2017 reach 16 % adoption of fixed broadband with at least 2 Mbps; by 2015 offer 20 Mbps to all medium and large enterprises
Czech Rep.	By 2013, in all populated localities a minimum of 2 Mbps and in cities a minimum of 10 Mbps. By 2015, rural areas to have at least one half of the average speed in cities and 30 % of premises in cities to have access to at least 30 Mbps
Denmark	By 2020 100 % of households and businesses to have access to 100 Mbps
Ecuador	By 2016, 90 % of Ecuadorian households and 95 % of large and medium enterprises to have access to at least 2 Mbps (residences), and 20 Mbps (enterprises)
Finland	By 2010, every permanent residence and permanent office of business or public administration body must have access to a fixed or wireless subscriber connection with an average downstream rate of at least 1 Mbit/s. By 2015 more than 99 % of population permanent residences and permanent offices of businesses or public administration bodies will be no more than 2 km from an optical fiber or cable network permitting 100 Mbps connections
France	By 2012, 100 % of the population to have access to broadband. By 2025 100 % of home to have access to very high speed broadband
Germany	By 2014, 75 % of households will have download speeds of 50 Mbps
Greece	By 2017, 100 Mbps to all homes
Hungary	By 2013, broadband coverage will be 100 %, and average speed will be 2 Mbps, with a target for 2020 of 30 Mbps
Iceland	2007: All Icelanders who so desire should have access to a high speed connection
India	By 2010, to have 20 million broadband connections
Ireland	October 2010: in areas where there was no broadband a mobile service (using HSPA), was required to be in place with a minimum download speed of 1.2 Mbps and a minimum upload speed of 200 kbps
Israel	Broadband included in universal service
Italy	By 2012, all Italians to have access to the Internet at between 2 and 20 Mbps
Japan	By 2015, fiber optic highways will be completed enabling every household to enjoy a broadband service

(continued)

Table 8.1 (continued)

Country	Targets
Korea	By 2010, to provide broadband multi-media services to 12 million households and 23 million wireless subscribers. By 2012 to raise average speeds to 10 Mbps with a maximum of 1 Gbps
Luxembourg	By 2015, FTTH to every household. By 2020 1, Gbps to every household
Mexico	By 2012, 22 % broadband penetration
New Zealand	By 2019, ultra fast broadband to 75 % of New Zealanders where they live, work and study. By 2015, 80 % of rural households to have speeds of at least 5Mbps, with the remainder to achieve speeds of at least 1 Mbps
Norway	By 2007, all citizens to be offered high speed broadband
Poland	By 2013, 23 % of population to have access to broadband. A citizen who has no computer may use one of the numerous points of access to digital services, which are located in public institutions
Portugal	By 2012, 100 % of municipalities covered by fixed NGN. By 2015, 100 % national coverage by LTE
Russia	By 2010, to have 15 lines per 100 population. By 2015, to have 35 lines per 100 population
Slovak Republic	By 2013, 100 % of population to have a minimum speed of 1 Mbps. By 2020, to provide access to high speed broadband of at least 30 Mbps
South Africa	By 2014, to have 5 % broadband penetration (min. 256 kbps)
Spain	By 2011, minimum speed of 1 Mbps broadband access available to 100 % of population. By 2015, 100 Mbps broadband available to 50 % of population
Sweden	By 2015 40 % of households and businesses should have access to 100 Mbps. By 2020 90 % of households and businesses should have access to 100 Mbps
Switzerland	Since 2008 a universal service obligation of 600 Kbps
United Kingdom	By 2015, to bring "superfast broadband" to all parts of the UK and to create the "best broadband network" in Europe. To provide everyone with at least 2 Mbps and superfast broadband to be available to 90 % of people
United States	By 2010, at least 100 million homes should have affordable access to actual download speeds of at least 100 Mbps and actual upload speeds of at least 50 Mbps. By 2020, every household should have access to actual download speeds of 4 Mbps and actual upload speeds of 1 Mbps

Source OECD. National Broadband Plans

population and economic and administrative units tend to overlap. Accordingly, if the objective is to cover 100 % of the population, it is highly likely that this goal will address 100 % of firms and public institutions.

In general term, coverage target setting is conducted a priori on the basis of social policy imperatives. The notion that broadband is a public service that needs to be provided to the whole population of a given country is gaining currency. Forty countries have so far stipulated in their policies and regulatory frameworks the need to provide universal access to broadband services (among them, Brazil, China, and Spain). Some countries go even further declaring broadband a human right (e.g., Finland, Estonia). Broadband plans of developed countries tend to define coverage targets for 100 % of population, while many emerging countries propose reaching 75 %.

Table 8.2 Matrix for defining national broadband planning targets

		Type of target		
		Supply (objective; network coverage)	Demand (objective: broadband adoption)	Speed of service (objective: Mbps of download)
Perspective	Residential (objective: universal service)	Percentage of population	Percentage of population[1]	10 Mbps of minimum download speed for universal service
	Social (objective: social inclusion)	Percentage of educational, cultural, health and scientific establishments	Percentage of educational, cultural, health and scientific establishments	10 Mbps of download speed per establishment
	Enterprises (objective: maximization of economic impact)	Percentage of industrial establishments	Percentage of large, medium, small and micro enterprises	10 Mbps of download speed per establishment and industrial sector
	Public administration (objective: maximization of efficiency)	Percentage of public administration units (offices)	Percentage of public administration units (offices)	10 Mbps of download speed per public administration unit

[1] Targets for supply and demand should be defined not only by the unit of population but also by the unit of household. Given that in many cases targets could be defined from a technology neutral position, broadband deployment can be achieved with either wired or wireless broadband. In that sense, measuring targets in terms of both metrics allows for better measurement and tracking. I think wired broadband can be chosen in several situations, such as urban area

Fig. 8.1 Ecuador: Historical and extrapolated broadband penetration

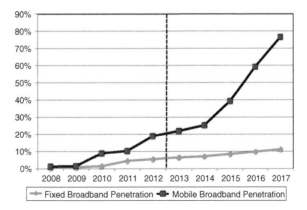

Residential demand targets Once coverage is established, demand targets need to be defined. This requires implementing different approaches for residential, social, enterprise, and public administration. In the case of residential demand targets, four methodologies could be used:

- Extrapolation of historical growth in penetration rates: this approach allows grounding residential target setting on a realistic base; however, it is important to consider potential saturation points when implementing this approach. For example, due to its late start, mobile broadband has been growing quite significantly in the past 2 years. Consequently, an extrapolation of penetration growth rate of this technology could overestimate future adoption (see example in Fig. 8.1).
- Correlation between economic development and broadband penetration: this methodology is based on a simple approach of determining whether the country under analysis is falling behind in broadband penetration when considering its level of economic development; target setting is defined in terms of penetration to be achieved in order to reach the level expected by its economic development level (see example in Fig. 8.2).
- Comparison with neighboring nations: this approach, while risking underestimated the target, a target that aligns the nation under consideration to regional penetration levels ("catch up approach").
- Country vision: an alternative approach to residential demand target setting stipulates that, based on broadband economic impact, the target should aim to grow its penetration in order to maximize its economic impact; this could be modeled by relying on econometric models of economic impact and assuming alternative impact scenarios. In other words, this methodology would establish a residential penetration target if the objective is to contributes by 10 % to growth of the GDP (see Fig. 8.3).

These four methodologies allow defining, by triangulation and validation, a consensus residential target.

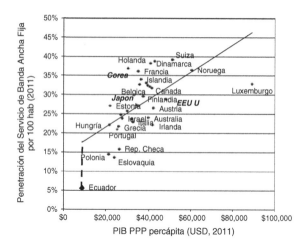

Fig. 8.2 Latin America: Economic development and Broadband penetration *Sources* World Bank; ITU; Telecom Advisory Services analysis

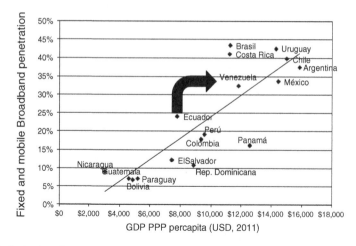

Fig. 8.3 Latin America: Economic development and Broadband penetration *Sources* Wireless Intelligence; ITU; World Bank; TAS analysis

Social demand targets the determination of targets for educational, cultural, research, and health institutions can be conducted by relying on the recommendations set up by International Telecommunication Union (see Table 8.3).

Enterprise demand targets the determination of broadband targets in the enterprise sector requires segmenting the sector by size of company and industrial sector. This segmentation is particularly relevant in the SME sector since there are firms that, despite their small size, are ICT intensive and require good broadband connectivity to fulfill their objectives. On the other hand, there are others that, due to

Table 8.3 Specific broadband penetration targets for social institutions

Type of institution	Type of target	Targets (%)
Universities, secondary and primary schools	Percentage of universities connected to high speed broadband	100
	Percentage of secondary schools connected to high speed broadband	100
	Percentage of primary schools connected to high speed broadband	100
ICT enabled scientific research centers	Percentage of research centers connected to high speed broadband	100
	Deployment of a national ultra-broadband network in support of scientific research	Yes
	Percentage of universities connected to national ultra-broadband network	100
	Percentage of scientific research centers connected to the national ultra-broadband network	100
Libraries and cultural centers equipped with ICT	Percentage of public libraries with access to the Internet	100
	Percentage of cultural centers with access to broadband	80
Hospitals and health centers	Percentage of hospitals connected to ultra-broadband	100
	Percentage of health centers connected to broadband	80

their position in low value-added domestic segments, are not necessarily in need of broadband.

This differentiation leads to the determination of different adoption targets by firm size and industrial sector.

Download speed targets download speed targets have to define by segment. While the minimum speed for universal residential broadband is one target, goals for social and enterprise applications will need to be differentiated by segment. For example, the targeted speeds for social institutions are driven by the broadband applications and their bandwidth intensity (see Table 8.4).

In the case of enterprises, it is useful to segment speed requirements by SME segment (see Table 8.5).

After defining all targets, national broadband plans could construct a matrix such as the one below, developed for Ecuador's National Broadband Plan (see Table 8.6).

8.2 Identifying and Managing Demand Stimulation Programs

Once targets are determined, national broadband plans have to identify specific policies to stimulate demand and bridge the gap. Whether explicit or implicit, there appears to be a link between broadband demand stimulation programs and national

Table 8.4 Target speeds for social institutions

Segments	Download speed
Universities	20–100 Mbps symmetrical
Primary schools	6–20 Mbps download
Research centers	20–100 Mbps symmetrical
Public libraries	6–20 Mbps download
Cultural centers	6–20 Mbps download
Hospitals	20–100 Mbps symmetrical
Health centers	6–20 Mbps download

Table 8.5 Target speeds for SMEs

20 Mbps symmetrical	2 Mbps download speed
• SMEs with capacity to access domestic and international markets, based on the supply of high value-added products or on the leveraging of a significant competitive advantage	• SMEs operating autonomously in low-ICT intensive sectors (retail trade, foodstuffs, chemical products, etc.)
• SMEs with a position in the value chain that leads them to supply inputs to large corporations	• Microenterprises in labor intensive sectors with low value added

strategies focused on broadband or digitization. A large portion of regional European Union digital literacy initiatives can be traced back to national strategies, such as the United Kingdom's "Empowering Citizens, Connecting Communities," or "Implementing the Information Society in Ireland," as well as regional plans, like the Lisbon Agenda and the Riga Declaration.

A classical country case in point is that of Finland, where the document "Finland—Towards an Information Society—A National Outline 1995–2000" served as a framework to defining a host of digital literacy initiatives. In addition, other policy initiatives formulated at a later date, such as the "Governmental Information Society program, 2003–2007" put additional emphasis on the deployment of initiatives aimed at promoting usability, inclusion and quality of life.

In the linkage between demand stimulation initiatives and national broadband plans, two differing philosophies exist. Some countries tend to deploy national and regional initiatives that are centrally managed and that have been defined in the national plans. In general, national centrally managed demand stimulation initiatives tend to originate within either the Ministry of Education or the Ministry/Secretariat for ICT/Communication. In their review of digital literacy initiatives, Hilding-Hannan et al. (2009) conclude that most initiatives, especially those led by public organizations, are generally driven by centralized policies at the national or regional level and are seen as strategically linked to government objectives. In particular, many initiatives and the rationales behind them can be traced back to either the economic or the social ramifications of a developing information society and government's priorities in response to these changing circumstances. A critical

Table 8.6 Ecuador: National Broadband Plan Targets

Perspective	Metas		
	Supply (objetive: network coverage)	Demand (objective: technology adoption)	Service speed (in Mbps)
Residential (objective: universal service)	• 100 % of population (2016)	• Fixed broadband (% population): • 11 % (2015) • 20 % (2017) • Mobile Broadband (% population): • 38 % (2015) • 63 % (2017)	• Fixed Broadband: • 2 Mbps download (2014)
Social (objective: social inclusion)	• 100 % educational, cultural, research and health establishments (2015)	• 100 % educational, cultural, research and health establishments (2015)	• 6–20 Mbps download (primary schools, public libraries, and health centers) • 20–100 Mbps symmetrical (hospitals, universities, and research centers)
Enterprises (objective: maximization of economic impact)	• 100 % enterprises (2015)	• 100 % large and medium enterprises (2013) • 100 % of ICT intensive SMEs (2015)	• 20 Mbps symmetrical (large, medium enterprises and SMEs in strategic sectors) • 2 Mbps download for microenterprises (2014)

advantage of national centralized programs is that they can tackle potential synergies and overlaps among demand stimulation initiatives.

In other nations, it is common for some initiatives to emerge at the local level, once the guidelines have been formulated nationally. Local initiatives are initiated at subsovereign levels (province, department, district, and even, municipalities). These local initiatives have an active participation of nongovernmental organizations and community interest groups. In that regard, the key advantage of locally driven programs has to do with their closeness with the communities being targeted. This feature allows the program to carefully tailor content and implementation dynamics to the needs of the targeted population. As a counter-example, a digital literacy program targeting SMEs in rural Canada was criticized for emphasizing training in presentation (e.g., PowerPoint) applications, not necessarily a critical need of participants.

A critical element in defining a successful demand stimulation program has to do with its long-term sustainability. In fact, according to Hilding-Hannan et al. (2009) sustainability remains a big concern, since according to their study of 464 programs across 32 countries, only 40 % have been identified as still running, transferred or expanded beyond the initial timeframe. Similar concerns were raised in the implementation of a municipal Wi-Fi program in Brazil. After installing the hardware, financing difficulties emerged in covering the ongoing operating costs of the facility. This has to do with the ability to implement the initiative until its objectives are accomplished. Hilding-Hannan et al. (2009) have identified five factors that are critical in driving program sustainability:

- Composition of stakeholders involved: the findings in this regard are quite clear. Digital literacy programs with higher chance of sustainability are either sponsored by a private company, the partnership of an NGO and a private company, a public-private partnership, or a public-private partnership including an NGO; on the other hand, if a project is exclusively sponsored by the public sector, chances are that it will not be sustainable. An obvious explanation of the linkage between private sponsorship and sustainability is the financial and infrastructure support, although the private entity likely focuses more on achieving the program objectives.

  In addition, in demand programs that are centrally managed but locally implemented, the number of stakeholders increases substantially, thereby raising the likelihood of program sustainability. Finally, the reliance on NGOs (local associations, community groups, voluntary organizations) drastically raises the probability of long-term sustainability because of the intrinsic interest of such entities in improving the livelihood of the population that lacks broadband access.

- Level of implementation: national programs, driven by a country policy or strategy appear to have a higher likelihood of sustainability (although the difference with regional and local programs is not that significant);

- Whether the initiative requires payment or not by the user: 60 % of analyzed digital literacy campaigns are free of charge to the users; and

- Whether the initiative is being evaluated as part of a larger project: while being a positive contributor to the program success, research indicates that evaluation is only conducted in the context of large-scale national initiatives.

Reference

Hilding-Hamann, K., Nielsen, M., Pedersen, K.: Supporting Digital Literacy: Public Policies and Stakeholders' Initiatives. [report] Copenhagen: The European Commission on Digital Inclusion (2009)

Conclusion

This book has addressed the multiple strategies, programs, and initiatives aimed at stimulating the adoption of broadband. It begun by recognizing that one of the chief obstacles to reaching broadband universalization was less focused on the "supply side" and more on addressing the "demand gap." It first proceeded by defining the demand gap as the population that could acquire broadband service because its being offered and yet they do not.

The book introduced the reader to the methodologies geared to building an understanding of the demand gap, through measurement and decomposition of the barriers of adoption. Beyond building the descriptive statistics that provide a view of the different dimensions of the demand gap by region of a given country, it is critical to understand the specific barriers: is it the demand gap existing because a portion of the population cannot afford to purchase a subscription at current prices? Or is it because they lack the necessary digital literacy that allows them to access the Internet? It could also be the case that, while potential users have a computer (or comparable device), they cannot find any online content, applications, or services that would motivate them to purchase broadband service.

Once the approaches to generating this basic understanding are presented, the book examined each of the three residential adoption barriers: limited affordability, limited awareness or lack of digital literacy, and lack of relevance of content. Similarly, it examined the barriers of Small and Medium Enterprise adoption: limited affordability, lack of technology training, and limited assimilation of broadband in the enterprise operational processes and functions. On this basis, the study provided a set of recommendations of best practices, potential strategies and case studies aimed at promoting broadband demand. The programs and practices were analyzed by adoption obstacle to be targeted.

The awareness and digital literacy barrier can be addressed through both embedding programs in the formal education system (primary, secondary students, and teachers) and targeting nonformal initiatives to specific segments of the population (adults, women, elderly, handicapped, rural poor, etc.). Best practices indicate that the structuring of digital literacy efforts should be conducted after concluding the basic diagnostic of demand gap. These initiatives should be complemented with the deployment of community access centers. In some cases, broadband access centers are installed in existing facilities as a complement of

R. L. Katz and T. A. Berry, *Driving Demand for Broadband Networks and Services*,
Signals and Communication Technology, DOI: 10.1007/978-3-319-07197-8,
© Springer International Publishing Switzerland 2014

activities already performed at those locations (e.g., public libraries). Other practices point to the creation of standalone access centers exclusively focused in providing free access to broadband services.

As a complement to the digital literacy programs, countries should consider initiatives regarding advanced ICT training, focusing on workforce development, development of capabilities of SME personnel, and creating awareness of the potential of broadband among government employees. Advanced ICT training can allow individuals to establish a career in the ICT industry, which typically offers higher quality and higher paying job opportunities. As such, this type of training ultimately allows a country to shift away from an economy dependent on manufacturing to one based on high value skills. Further, advanced ICT skills can permeate other industries, improving business practices in, for example, the health care and finance sectors.

Beyond awareness and digital literacy, research indicates that limited affordability is a critical adoption obstacle when broadband penetration ranges between 3 and 20 % of the population, which is the stage at which most emerging countries are. The book first presented the economics of broadband adoption by introducing all the components driving the total cost of ownership of the technology. They comprise device acquisition and other one-time costs, service subscription retail pricing (with multiple sub-components), and service taxation. This introduction served as a basis to discuss the potential policy and private sector initiatives addressing the broadband affordability obstacle.

Three types of initiatives, targeting the affordability obstacle were examined. Service pricing was discussed for both fixed and mobile broadband. It introduced an approach for conducting comparative pricing analysis, followed by presenting models of service price elasticity, ending with a review of potential private sector and public policy initiatives to reduce service pricing. Next, the study turned to device pricing. Broadband access requires devices capable of accessing the Internet. They range from computers supplemented with a modem (called USB modem, dongle, or air card) to smartphones, netbooks, and tablets. Since pricing dynamics (and capabilities) vary greatly across devices, the demand structural factors linked to device access were discussed in two distinct sections: personal computers and mobile devices. In this context, several potential programs aimed at reducing device pricing were presented. The final set of affordability initiatives focused on taxation. This area reviewed the different equipment levies (import duties, value added taxes, and sector specific) and service taxes (value added and sector specific). This review served as a basis to discuss the impact of taxation on total cost of ownership of broadband, and present examples of policy initiatives to tackle broadband taxation.

After dealing with the policy initiatives aimed at increasing affordability, the study turned to how to overcome one of the three dominant adoption obstacles to broadband diffusion among residential subscribers: lack of content relevance. Research indicates that, at higher penetration levels of broadband (beyond 20 %), price elasticity coefficients start to decline, signaling that affordability plays a smaller role in constraining diffusion. At that point, lack of relevant content remains the final obstacle to achieving universal adoption. Demand stimulation centers here on enhancing broadband's value proposition. This is contingent upon offering applications and content that build service attractiveness. The study presented the multiple dimensions of content relevance, ranging from the linguistic to the cultural and applications dimensions.

The study concluded by examining the potential social and economic return of enhancing broadband adoption and introducing methodologies contributing to focusing on demand stimulation in the development of national broadband plans.

Building on the evidence and best practices from past programs, this study attempted to provide a holistic tool and guide regulators and policy makers in addressing broadband demand obstacles. We hope that, by shedding some light on this critical issue, this study contributes to bridging the digital divide and serve as an impetus for further research.

About the Authors

Dr. Raul Katz is President of Telecom Advisory Services LLC, a consulting firm active in the fields of strategy, economic analysis, and regulation. He previously served as a Lead Partner at Booz Allen Hamilton, where he was a member of the firm's Leadership Team and Head of the US and Latin America telecommunications practices. As an international telecom industry consulting executive, Dr. Katz has provided direction to top management of major telecommunications, software, and information services companies, focusing on business strategy, consumer/industrial marketing, and general management approaches. In addition, Dr. Katz has worked with governments and international organizations to develop regulatory frameworks and policies, National Broadband Plans, Digital Agendas, and National Technology Strategies. In particular, he has supported the governments of Colombia, Peru, Costa Rica, Ecuador, and Mexico.

Dr. Katz also serves as the Director of Business Strategy Research at Columbia University's Center for Tele-Information, is an Adjunct Professor in Columbia Business School's Division of Finance and Economics, and a Visiting Professor at the Universidad de San Andres in Argentina. He has published articles in journals such as *Telecommunications Policy*, *Technovation*, *Info*, *Strategy and Business*, *Communications and Strategies*, *Intereconomics*, and *America's Network*. His first book, *The Information Society: an International Perspective,* which focuses on the deregulation trends in the worldwide telecommunications industry, was published in 1988. His second book, *Creative Destruction: Business Survival Strategies in the Global Internet Economy*, which addresses discontinuities in the telecommunications industry, was published in 2000 and translated into Japanese. His third book, *The Role of ICT in Development,* was published in 2010.

He holds a Ph.D. in Management Science and Political Science and a MS in Communications Technology and Policy from MIT, a Maitrise and a Licence in Communications Sciences from the University of Paris, a Maitrise in Political Science and a Licence in History from the Sorbonne.

Taylor Berry has served as a consultant with Telecom Advisory Services since 2011 and first collaborated with Dr. Katz at Columbia University's Center for Tele-Information. With a background in communication technology, Ms. Berry's

R. L. Katz and T. A. Berry, *Driving Demand for Broadband Networks and Services,*   311
Signals and Communication Technology, DOI: 10.1007/978-3-319-07197-8,
© Springer International Publishing Switzerland 2014

professional experience has centered on Internet industry and global telecommunications trends, with particular attention devoted to the impact of increased access to these services. Ms. Berry received her Master of Arts from Georgetown University, where she completed a dual-degree program in Communication Technology and International Business Diplomacy. Prior to attending Georgetown, Ms. Berry studied at the Graduate Institute of International and Development Studies in Geneva, Switzerland and received her Bachelor of Arts from Wake Forest University.

CPSIA information can be obtained at www.ICGtesting.com
Printed in the USA
BVOW10*2239140714

359197BV00001B/2/P